AF081939

Creating a More Transparent Internet

On social media, new forms of communication arise rapidly, many of which are intense, dispersed, and create new communities at a global scale. Such communities can act as distinct information bubbles with their own perspective on the world, and it is difficult for people to find and monitor all these perspectives and relate the different claims made. Within this digital jungle of perspectives on truth, it is difficult to make informed decisions on important things like vaccinations, democracy, and climate change. Understanding and modeling this phenomenon in its full complexity requires an interdisciplinary approach, utilizing the ample data provided by digital communication to offer new insights and opportunities. This book gives a comprehensive view on social media communication, the different forms it takes, the impact and the technology used to mine it, and defines the roadmap to a more transparent Web. This interdisciplinary approach will be of interest to researchers in computational linguistics, linguistics, social sciences, media studies, the Semantic Web, and internet technology.

PIEK VOSSEN is Professor in Computational Linguistics at the Vrije Universiteit Amsterdam. He is Co-founder and President of the Global Wordnet Association and has received the Dutch Spinoza Prize in 2013 for his research. He used this prize for projects on language understanding, including mining perspectives in debates like on vaccination.

ANTSKE FOKKENS is Professor in Computational Linguistics at the Vrije Universiteit Amsterdam and Associate Professor at the Applied Geometric Algorithms group at the Technical University Eindhoven. Since 2021, she has acted as University Research Chair on methodological aspects of computational linguistics, in particular when used in an interdisciplinary setting.

STUDIES IN NATURAL LANGUAGE PROCESSING

Volumes in the SNLP series provide comprehensive surveys of current research topics and applications in the field of natural language processing (NLP) that shed light on language technology, language cognition, language and society, and linguistics. The increased availability of language corpora and digital media, as well as advances in computer technology and data sciences, has led to important new findings in the field. Widespread applications include voice-activated interfaces, translation, search engine optimization, and affective computing. NLP also has applications in areas such as knowledge engineering, language learning, digital humanities, corpus linguistics, and textual analysis. These volumes will be of interest to researchers and graduate students working in NLP and other fields related to the processing of language and knowledge.

Also in the series

Douglas E. Appelt, *Planning English Sentences*
Madeleine Bates and Ralph M. Weischedel (eds.), *Challenges in Natural Language Processing*
Steven Bird, *Computational Phonology*
Peter Bosch and Rob van der Sandt, *Focus*
Pierette Bouillon and Federica Busa (eds.), *Inheritance, Defaults and the Lexicon*
Ronald Cole, Joseph Mariani, Hans Uszkoreit, Giovanni Varile, Annie Zaenen, Antonio Zampolli, and Victor Zue (eds.), *Survey of the State of the Art in Human Language Technology*
David R. Dowty, Lauri Karttunen, and Arnold M. Zwicky (eds.), *Natural Language Parsing*
Ralph Grishman, *Computational Linguistics*
Graeme Hirst, *Semantic Interpretation and the Resolution of Ambiguity*
András Kornai, *Extended Finite State Models of Language*
Kathleen R. McKeown, *Text Generation*
Martha Stone Palmer, *Semantic Processing for Finite Domains*
Terry Patten, *Systemic Text Generation as Problem Solving*
Ehud Reiter and Robert Dale, *Building Natural Language Generation Systems*
Manny Rayner, David Carter, Pierette Bouillon, Vassilis Digalakis, and Matis Wiren (eds.), *The Spoken Language Translator*
Michael Rosner and Roderick Johnson (eds.), *Computational Lexical Semantics*
Richard Sproat, *A Computational Theory of Writing Systems*
George Anton Kiraz, *Computational Nonlinear Morphology*
Nicholas Asher and Alex Lascarides, *Logics of Conversation*
Margaret Masterman (edited by Yorick Wilks), *Language, Cohesion and Form*
Walter Daelemans and Antal van den Bosch, *Memory-Based Language Processing*
Chu-Ren Huang, Nicoletta Calzolari, Aldo Gangemi, Alessandro Lenci, Alessandro Oltramari, and Laurent Prévot (eds.), *Ontology and the Lexicon: A Natural Language Processing Perspective*
Thierry Poibeau and Aline Villavicencio (eds.), *Language, Cognition, and Computational Models*
Bind Liu, *Sentiment Analysis: Mining Opinions, Sentiments, and Emotions, Second Edition*
Marcos Zampieri and Preslav Nakov, *Similar Languages, Varieties, and Dialects: A Computational Perspective*

Creating a More Transparent Internet

The Perspective Web

Edited by

PIEK VOSSEN
VU University Amsterdam

ANTSKE FOKKENS
VU University Amsterdam

Section Editors
CAMIEL BEUKEBOOM
VU University Amsterdam

JULIA NOORDEGRAAF
VU University Amsterdam

THOMAS POELL
VU University Amsterdam

IVAR VERMEULEN
VU University Amsterdam

CAMBRIDGE
UNIVERSITY PRESS

University Printing House, Cambridge CB2 8BS, United Kingdom

One Liberty Plaza, 20th Floor, New York, NY 10006, USA

477 Williamstown Road, Port Melbourne, VIC 3207, Australia

314-321, 3rd Floor, Plot 3, Splendor Forum, Jasola District Centre, New Delhi – 110025, India

103 Penang Road, #05–06/07, Visioncrest Commercial, Singapore 238467

Cambridge University Press is part of the University of Cambridge.

It furthers the University's mission by disseminating knowledge in the pursuit of education, learning, and research at the highest international levels of excellence.

www.cambridge.org
Information on this title: www.cambridge.org/9781108485760
DOI: 10.1017/9781108641104

First published 2022

A catalogue record for this publication is available from the British Library

Library of Congress Cataloging in Publication data
Names: Vossen, Piek, editor. | Fokkens, Antske, 1982- editor.
Title: Creating a more transparent Internet : the Perspective Web / edited by Piek Vossen and Antske Fokkens.
Description: Cambridge, United Kingdom ; New York, NY : Cambridge University Press, [2022] | Series: SNIP studies in natural language processing | Includes bibliographical references and index.
Identifiers: LCCN 2021058280 (print) | LCCN 2021058281 (ebook) | ISBN 9781108485760 (hardback) | ISBN 9781108641104 (ebook)
Subjects: LCSH: Natural language processing (Computer science) | Semantic computing. | Perspective (Linguistics) | Internet. | BISAC: LANGUAGE ARTS & DISCIPLINES / Linguistics / General
Classification: LCC QA76.9.N38 C74 2022 (print) | LCC QA76.9.N38 (ebook) | DDC 006.3/5–dc23/eng/20220105
LC record available at https://lccn.loc.gov/2021058280
LC ebook record available at https://lccn.loc.gov/2021058281

ISBN 978-1-108-48576-0 Hardback

Contents

List of Contributors *page* x

1 **Introducing the Perspective Web** ***Piek Vossen and Antske Fokkens*** 1

1.1 A Web of Perspectives 1
1.2 The Urgency of the Perspective Web 2
1.3 What This Book Is About 3
1.4 The Structure of This Book 5
References 6

PART I THEORETICAL BACKGROUND
Section Editors: *Piek Vossen and Antske Fokkens*

2 **Perspectives from a Social Psychological and Communication Scientific Perspective** ***Ivar Vermeulen and Camiel Beukeboom*** 9

2.1 Introduction: Perspectives as (Embedded) Belief and Attitude Structures 9
2.2 Recipient Perspectives: The Role of Perspectives in Processing Incoming Information 12
2.3 Sender's Perspectives: How a Sender's Perspective is Reflected in Communication and Language Use 15
2.4 Social and Interactive Dynamics 18
2.5 Conclusion 25
References 26

3 **Computational Linguistics for Subjectivity** ***Preslav Nakov*** 31

3.1 Sentiment Analysis 31
3.2 Other Subjectivity Tasks 36

3.3 Machine Learning Models 39
3.4 Conclusion and Recommended Readings 43
References 44

PART II SOCIAL IMPACT
Section Editors: *Ivar Vermeulen and Camiel Beukeboom*

4 Perspectives in a Social Context: The Role of Communication ***Ivar Vermeulen and Camiel Beukeboom*** 57

5 The Micro Level: Linguistic Perspective in Written Discourse ***Kobie van Krieken and José Sanders*** 60
5.1 Perspective in Language and Discourse 60
5.2 A Text-Linguistic Approach to Perspective in Narrative 62
5.3 Focalization 64
5.4 Demonstration versus Invasion 65
5.5 Quotations 66
5.6 Direct Speech 68
5.7 Indirect Speech 68
5.8 Free Indirect Speech 69
5.9 Multiple Perspectives 70
References 71
5.10 Appendix: I Sympathise with You Guys, Lots of Luck 73

6 The Meso Level: Perspectives in a Social Context ***Rachel Neo*** 75
6.1 The Meso Level 75
6.2 Theories of Computer-Mediated Communication and New Media 75
6.3 Traditional Theories of Media Effects 82
6.4 Conclusion 86
References 88

7 The Macro Level: Perspectives Embedded in Society, Culture, and Technology ***Hong Vu*** 94
7.1 Introduction 94
7.2 Technologies for Personalized Content 95
7.3 Selective Exposure 98
7.4 Homophily, Echo Chamber, and Filter Bubble 100
7.5 Political Polarization and Audience Fragmentation 104
7.6 Concluding Remarks 105
References 106

PART III MEDIATING PERSPECTIVES
Section Editors: *Julia Noordegraaf and Thomas Poell*

8 **The Mediation of Online Information** ***Julia Noordegraaf and Thomas Poell*** 113
References 117

9 **The Source and Its Encoding: Reflections on Metadata in Digitized and Born-Digital Media Collections** ***Eric Hoyt*** 119
9.1 Introduction 119
9.2 Metadata in Digitized Magazine Collections 121
9.3 Podcasts, RSS Feeds, and the Encoding of Born-Digital Objects 124
9.4 Conclusion 127
References 128

10 **Knowledge-Making on Techno-Commercial Platforms: The Example of Facebook** ***Jonas Anderson Schwarz*** 129
10.1 Introduction 129
10.2 The Normative Power of Prescribed Social Action 130
10.3 The Corporate Mission 133
10.4 Objectives and Ethics 136
10.5 Conclusion: Perspective-Making on, of, and through Facebook 138
References 140

11 **Content, Form, and Reception: Perspectives from Digital Media Data** ***Christina Neumayer*** 143
11.1 Introduction 143
11.2 A Brief Note on Perspectives from Political Protest 144
11.3 Content 145
11.4 Form 147
11.5 Reception 149
11.6 Perspectives from Digital Data 151
References 153

12 **Quality and Perspectives** ***Davide Ceolin, Julia Noordegraaf, and Lora Aroyo*** 156
12.1 Introduction 156
12.2 Information Quality Dimensions 158
12.3 Related Work 160
12.4 A Tool for Information Quality Assessment 161
12.5 Conclusion and Future Work 166
References 167

PART IV MINING AND MODELING PERSPECTIVES

Section Editors: *Piek Vossen and Antske Fokkens*

13 Mining and Modeling Perspectives ***Piek Vossen and Antske Fokkens*** 171

14 Natural Language Processing Tasks for the Extraction of Perspectives ***Chantal van Son, Roser Morante, and Piek Vossen*** 173
14.1 Introduction 173
14.2 Modeling and Processing Attribution 176
14.3 Factuality Profiling 178
14.4 Natural Language Inference 182
14.5 Micro-Propositions 184
14.6 Conclusions 187
References 188

15 Toward Automatic Discovery of Diverse Perspectives ***Sihao Chen, Daniel Khashabi, and Dan Roth*** 193
15.1 Introduction 193
15.2 Minimal Perspective Discovery: Tasks and Challenges 195
15.3 Related Work 197
15.4 The *Perspectrum* Dataset 198
15.5 Building Systems Based on *Perspectrum* 199
15.6 Toward Automated Perspective Discovery 203
15.7 Discussion and Conclusion 204
References 205

16 Formal Representation and Extraction of Perspectives ***Aldo Gangemi and Valentina Presutti*** 208
16.1 Viewpoints and Perspectives: A Landscape 208
16.2 What is a Perspective? 211
16.3 Representing and Extracting Perspectives 213
16.4 Conclusions 225
References 226

17 The User Perspective in Professional Information Search ***Suzan Verberne*** 229
17.1 Introduction 229
17.2 Professional Search 231
17.3 Personalized IR 232
17.4 Explainable Recommendation and Search 234
17.5 Toward Explainable Professional Search 235

17.6 Conclusions 238
References 239

18 Harvesting Perspectives in Social Media ***Tommaso Caselli and Malvina Nissim*** 244
18.1 Introduction 244
18.2 Methods 245
18.3 Case Studies 249
18.4 Conclusions 255
References 255

19 GRaSP: A Model for the Perspective Web ***Piek Vossen and Antske Fokkens*** 260
19.1 Introduction 260
19.2 A Model for the Perspective Web 262
19.3 The GRaSP Model 263
19.4 Related Work 269
19.5 A Future Perspective on the Web 271
References 275

Contributors

Lora Aroyo
Google Research NYC

Camiel Beukeboom
Vrije Universiteit Amsterdam

Tommaso Caselli
University of Groningen

Davide Ceolin
Centrum Wiskunde & Informatica, Amsterdam

Sihao Chen
University of Pennsylvania

Antske Fokkens
Vrije Universiteit Amsterdam

Aldo Gangemi
University of Bologna and ISTC-CNR, Rome

Eric Hoyt
University of Wisconsin-Madison

Daniel Khashabi
Allen Institute for Artificial Intelligence

Kobie van Krieken
Radboud University Nijmegen

Roser Morante
Vrije Universiteit Amsterdam

Preslav Nakov
Qatar Computing Research Institute, HBKU

Rachel Neo
University of Hawaii at Manoa

Christina Neumayer
University of Copenhagen

Malvina Nissim
University of Groningen

Julia Noordegraaf
University of Amsterdam

Thomas Poell
University of Amsterdam

Valentina Presutti
University of Bologna

Dan Roth
University of Pennsylvania

José Sanders
Radboud University Nijmegen

Jonas Anderson Schwarz
Södertön University Stockholm

Chantal van Son
Vrije Universiteit Amsterdam

Suzan Verberne
University of Leiden

Ivar Vermeulen
Vrije Universiteit Amsterdam

Piek Vossen
Vrije Universiteit Amsterdam

Hong Vu
University of Kansas

1
Introducing the Perspective Web

Piek Vossen and Antske Fokkens

1.1 A Web of Perspectives

It is the year 2021. The year in which the impact of digital communication resulted in a direct physical assault on the senate in Washington, DC, on January 6, now called the "Capitol Riots." An assault by people that, although spread out over the country, formed a community of believers that the presidential elections had been "stolen." Never before have we witnessed the impact of communication on our world of beliefs in such a direct and forceful way. How could this happen? How did such extreme polarization come about?

The obvious suspect here is the Internet that has revolutionized our social communication and the high-tech companies that operate it. In a decade, we changed from consumers to prosumers of information and media. As a community, we post massive amounts of information and data on the Web. Our communication has never been so intense, and it is not all gold that glitters. Hate speech, fake news, hoaxes, disinformation, misinformation, trolling, phishing, identity theft, digital bashing, among others, are all new concepts that were introduced by this new media. The digitalization of social communication has connected us but also divided us into new communities, each with their own perspective on the world. Truth has become an opinion among soul mates that meet each other in impermeable filter bubbles. The promise of the nineties that the Internet brings the information society failed. Instead of free access to knowledge and enlightenment, we find ourselves instead in a communication society with unsupported claims, lies, errors, deception, misquotes, twisting, insults, and threats. Through social media, we communicate more data in weeks back and forth than has been published in centuries. For sure, the Web is an enormous network of dynamic data. But what is this data, what does it represent, and can we understand this dynamic process with its digital traces? Can we use technology to record it, detect evolving communities and their identities, determine the content and form of communication and the impact? Can we distill precise communication models, explain what is observed, and even predict what will happen? These are questions that we try to address in this book. Despite the dystopian view of the Web, we cannot deny that it provides a unique opportunity for researchers with very different backgrounds to do concrete research on one of the most complex social phenomena of humanity.

1.2 The Urgency of the Perspective Web

Finally, technology created the Web but should technology also take responsibility and further develop it and support people that now partly live in a digital world? Let us consider a concrete example. The following text fragment was posted on a website laleva.org in 2015 by Brian Shilhavy, Health Impact News Editor:

According to a statement made by Dr. Anne Schuchat, the director of CDC's National Center for Immunization and Respiratory Diseases, in an Associated Press story picked up by Fox News on April 25, 2014: "There hs been no measles deaths reported in the U.S. since 2003." www.laleva.org/eng/2015/03/

The text contains a citation that takes the form of a simple statement and is attributed to an official from a US government agency. The statement itself looks like straightforward text, but there is no such thing as simple text. The publication of this quote is a complex message with many different layers of information. For a start, the quote contains a typo, which does not add to its trustworthiness. More seriously, it is partial and taken out of context. Dr. Schuchat is claiming the opposite of what is suggested in the original article in the Associated Press:

Associated Press, Author: Mike Stobbe, Published Thursday, April 24, 2014: "Wrapped into that estimate are 71 million measles cases, nearly 9 million measles hospitalizations and 57,000 measles deaths. There has been no measles deaths reported in the U.S. since 2003. 'But the way we're going, we feel it (another) is inevitable,' Schuchat said."

The statement itself is not so trivial. It is not clear in this sentence as such what the focus is: is it on the "no," on "U.S." or on "since 2003"? Without the context, this is difficult to tell. In the original text, it is clear how this statement is contrasted with the statistics at a global scale arguing that the United States is neglecting the threat. Besides neglecting the context of the quote, Brian Shilhavy clearly stresses the authority of his source both as a doctor and director of a government agency and also the authority of the medium being Associated

Press. The example shows the complexity of the communication and the slyness with which people use their skills to get the impact they want.

The vaccination debate in which this post takes part has gone on for centuries since it was first conceived. Today, during the COVID-19 pandemic, it is more actual than ever. It relates to a broader societal movement in which people do not trust the government, science, the media, and the industry (Big Pharma); they claim their right to choose their own truth with the effect that less and less people vaccinate their children. Information and knowledge are spread wildly, while passionate groups are formed through online media. The positions are relatively simple: vaccines are either good or bad, but the argumentation and information are emotional, obscure, wrong, and misleading. The language is rich, complex, and powerful, while the impact has big consequences. How can parents find the "right" information and make a decision on the vaccination of their child? How to trace and trust sources? How to separate emotion from logical arguments? How to detect false arguments? How to know what is out there, outside your bubble? How to find the stances and arguments and how to position all stakeholders around it?

We believe that researchers not only have the opportunity to analyze and study social interactions, language, communication, and their impact in the digital age, but they also have the obligation to use their insight, skills, and technology to reshape this digital media into a safer and more transparent medium. We therefore propose the notion of the Perspective Web. The Perspective Web can be seen as a function, like a night-vision, that reveals hidden traces of information and claims so that you can see who is behind them, who believes what, how it is authorized, how (un)certain sources are making their claims, how claims can be twisted, and who else shares these beliefs? Such a view on the Web and social media would make communication and information more transparent and trustworthy. The Perspective Web could help measure consistency and judge quality of information. Such a Perspective Web requires the development of technology that can digest the current jungle of information and overlay it with a structured map of argumentation, positions, and sources. Do we understand how we communicate in these online debates, which are dispersed and so dynamic? If communication is so complex and so much is unsaid, can we build software that can "read" the debate and understand it? Can technology build a Perspective Web?

1.3 What This Book Is About

In order to answer these questions, we need to bring together experts from a large variety of disciplines. This book is therefore about perspectives in many different ways. It addresses perspectives of people on people, information, and knowledge, as expressed on the Web. People's perspectives can be modeled and tested to explain people's behavior. However, this digital variant of social interaction and communication also requires that we analyze it from different scientific perspectives. The complexity of the communication and the complexity of the social and psychological impact go hand in hand. Shannon and Beaver (Shannon, 1948) were the first to propose a model for communication. Their mathematical model distinguishes a source sending a message, a transmitter that encodes a message into a signal, a channel through which it is sent and a receiver that decodes it at a destination. It is no surprise that they were engineers from the telephone company Bell. Shannon and Beaver described three success factors for communication: (1) how accurately can the message be transmitted, (2) how precisely is the meaning "conveyed," and (3) how effectively does the received meaning affect behavior? Their model was adapted by others (Berlo, 1960; Schramm, 1954) to model person-to-person communication by distinguishing: sender, message, form, channel, and receiver.

A classic Natural Language Processing approach would consider public debates as data, mostly in text form and sometimes as a discourse, for example, a discussion on a forum, news articles, or a Twitter stream. The textual units (words, phrases, sentences, etc.) are labeled through an annotation process and language models that represent these units are fine-tuned to learn the features to predict these annotations. These labels can be emotions, relations, or attributes such as fake and offensive. Such approaches focus on the text as a signal as such without considering the communicative context, the sender, the receiver, and in some cases not even the channel. A signal-only view on debates will have a hard time grasping the full complexity and interpreting the signal deeply. It also does not consider the social relationships expressed through the process of signaling, nor group identities or relationships across dispersed signals. Finally, it lacks data on the impact on the receiver and community effects at a large scale.

If we want to capture this process in its totality, we need to take a cross-disciplinary approach that addresses the participating people, the form of their posts and blogs, the content, and the diverse media through which they are distributed. Social-psychological research targets complex sender–receiver interactions, linguistic and computational approaches describe and model the message, its form, and its meaning, whereas media analysis also takes impact of the channel into account. The channel itself is complex in this modern digital media landscape. Not only is there a plethora of devices and applications, each with their specifics, we are also dealing with a handful of tech companies that control these channels, while the old media powers still hold on to their traditional rights on content. Meanwhile, we see politicians and governments trying to get to grips with abundant new developments hardly constrained by national borders, crafting laws and regulations of which we cannot see the effect or impact either.

1.4 The Structure of This Book

In this book, we use the idea of a Perspective Web to aggregate the state-of-the-art view from different disciplines so that we can put these works in contexts: Social Sciences, Media Studies, Computational Linguistics, and Semantic Web scientists, on a central theme with high societal relevance. The book[1] is divided into four parts. In Part I, we introduce the theoretical notions on: (1) the linguistics of argumentative language, (2) the computational linguistic approaches to model and capture this, and (3) the social-psychological concepts to understand communication. Next, we dive deeper into the matter in three subsequent parts. Part II reflects on the so-called micro, meso, and macro levels of communication impact. The micro level focuses on the written discourse, the meso level on the direct social environment, and the macro level on the larger cultural and technological factors. Whereas Part II mainly considers communication as a social-psychological phenomenon, Part III looks at the online mediation process. This mediation process is analyzed in different dimensions, ranging from source encodings, channel politics, and technicalities, to content, form, and perception. The final chapter discusses quality of interaction in a computational environment, taking a more applied perspective. Part IV of the book is more technical, as it discusses how to model structures and interpretation computationally. For a start, Natural Language Processing techniques can be used to detect various properties of posted texts, including their argumentative structure, their sources, and their views on claims. Such an analysis can be used to organize claims into camps and provide a more comprehensive and less biased overview. In addition to interpreting expressions in text, we also need to model these claims as beliefs of sources (people) in formal models that can be used to reason over the implications. Whereas text is a message or signal that can be repeated many times whether confirmed or denied, these formal models condense the content of the message to their semantics. Finally, Part IV ends with two chapters that discuss how Natural Language Processing can

[1] This book is the outcome of a Lorentz-workshop https://nias-lorentz.nl/previous-ssh-workshops/ that was organised in 2017. At this workshop, international researchers from many different disciplines discussed the idea of the Perspective Web.

be used within common applications we use daily in order to mitigate some of its negative impact. In the final chapter, we present a future perspective on Natural Language Processing as a core technology that will play a central role in various dystopic and utopic scenarios of the near future. We hope that this books inspires researchers to collaborate more across disciplines so that we not only have a better understanding of what is going on on social media and digitalized communication but also will be able to find solutions and answers to the deficits of this platform.

References

Berlo, D. K. 1960. *The process of communication.* New York: Holt, Rinehart, & Winston.

Schramm, W. 1965. How communication works. In *The process and effects of mass communication*, 6th ed. Edited by: Schramm, W. 3–26. Urbana: University of Illinois Press.

Shannon, C. E. 1948. A mathematical theory of communication. *The Bell System Technical Journal*, **27**(3), 379–423.

PART I

Theoretical Background

Section Editors: Piek Vossen and Antske Fokkens

2

Perspectives from a Social Psychological and Communication Scientific Perspective

Ivar Vermeulen and Camiel Beukeboom

2.1 Introduction: Perspectives as (Embedded) Belief and Attitude Structures

In this chapter, we discuss how perspectives have been approached in social psychology, communication science, and linguistics. Let us first discuss how perspectives are defined in these fields. From a psychological point of view, a perspective can be conceptualized as the existing knowledge, personally held views, beliefs, attitudes, or opinions about a given topic or object, as residing in an individual's mind. The most straightforward type of perspectives are beliefs and attitudes. Beliefs are bits of private knowledge about entities, such as "this guy works for Sony," or "beer mixes poorly with chocolate." Attitudes are basically affective evaluations of entities, ranging from positive to negative, such as "I don't like this guy," or "I love beer." The object of interest can also concern a broad entity, such as a generic social category of people (e.g., Germans, scientists). In these cases, the associated set of beliefs is termed "stereotype," while attitudes about social categories are termed "prejudice."

Beliefs and attitudes are relatively stable cognitive elements. We have learned them over time, and, once learned, they are somewhat resistant to change. The stronger beliefs and attitudes are, the harder they are to change, and the more predictive they are for behavior (Pomerantz et al., 1995). Depending on the context, belief and attitude strength may be operationalized as accessibility from memory, certainty, or personal involvement, or as a combination of these three. This implies, for example, that attitudes that are harder to access may be more easily adjusted and less predictive of behavior – and vice versa (Fazio, 1995).

An important reason why beliefs and attitudes are relatively stable is that they exist in closely connected associative networks in our minds (Dalege et al., 2016; Eagly and Chaiken, 1995). First, beliefs and attitudes are mutually connected. In most models, attitudes are based on underlying beliefs: we usually do not "just" like someone, we like a person because of things we know about them, such as perceiving the person as caring, handsome, or successful. However, attitudes may also be learned relatively independently from underlying beliefs, for example, by implicit processes such as mere exposure or conditioning, where we unconsciously learn to associate a particular entity with positive affect (Olson and Fazio, 2001). In such cases, it may be the case that our beliefs about the entity are fueled by our attitude about it, rather than the reverse; an example could be that some consumers may have learned to love the brand Nike because of the affective responses elicited by its ads, and for that reason believe that the quality of Nike's products is always high. Second, individual beliefs and attitudes are embedded in larger cognitive structures, such as norms and values, or political, cultural, or religious world views (Dalege et al., 2016). For example, people may feel strongly about values such as equality or freedom, or about national identity, environment, or religion. In general, we tend to bring our cognitive beliefs and attitudes in line with such broader worldviews. For example, we may dislike particular politicians or laws because they do not align with our views on equality. Or our religious or environmental views may prevent us from enjoying pork. When inconsistencies arise (either between cognitions or between simple cognitions and broader worldviews), we experience cognitive dissonance – a negative affective state inciting us to bring back consistency (Shultz and Lepper, 1996). As a result, individual cognitions (e.g., attitudes toward politicians or pork) will likely be adjusted to the larger worldviews; the latter are harder to adjust because they tend to consist of thousands of mutually consistent cognitions (Monroe and Read, 2008).

Third, beliefs and attitudes exist in persons with varying personality characteristics. People may, for example, be more or less sensitive to social norms (self-monitoring), more or less inclined to mull information over (need for cognition), more or less disposed to consider different points of view (dogmatism), and so on. Such differences in personality structures may result in the formation of quite different attitudes and beliefs on the basis of similar information (e.g., Haugtvedt and Petty, 1992). Additionally, people also are susceptible to moods – temporal affective states generally categorized by valence (from positive to negative) and arousal (from high to low arousal) – and emotions. These affective states may also alter how attitudes are formed: think about how stressed versus relaxed employees may respond to particular work-related communications (Petty and Briñol, 2015). And finally, of course, gender, age, cognitive capacities, socioeconomic status, knowledge, media use, substance use, important life events, and many other psychological, biological, social, economic, behavioral, technical, and contextual factors may mold perspectives, sometimes in unpredictable ways (Schwarz, 2007).

In sum, perspectives can be seen as the knowledge, beliefs, attitudes, and opinions we may have about very specific topics (e.g., what to eat tonight, a celebrity's haircut) and on very broad and global topics (e.g., the media industry, parenting). Yet, as such attitudes and beliefs are connected and embedded in larger structures of existing knowledge, a person's perspective can also be seen more broadly as the combination of a person's mutually connected cognitions, affective reactions, and personality tendencies resulting from one's cultural, social, and individual background, which comes into play when approaching new situations and information.

Moreover, it is important to note that our perspectives are situated in the social contexts in which they come about (Smith and Semin, 2004). When people interact and communicate with each other, they share their individual perspectives on a given topic, and this results in various dynamic interactive processes, which we will discuss in the subsequent sections. To analyze the social dynamics of perspectives in communication it helps to distinguish a sender and a recipient role. In any type of communication, one can distinguish a sender (i.e., an author, speaker) who communicates their perspective in some form (e.g., verbal or nonverbal, spoken, written language) to recipients (i.e., a reader, a listener, an audience). In interactive media (e.g., conversation), one will constantly switch between these roles.

Senders and recipients have their own individual perspectives on a given topic, and these perspectives can be congruent or incongruent with each other. Perspectives are not necessarily fixed. A person's perspective can change when one is confronted with an alternative perspective communicated by an interaction partner or in an article text. In fact, changing another person's perspective may sometimes be a goal when one tries to persuade or convince someone. However, rather than attempting to change a recipient's perspective, a sender may also adapt their utterances to an assumed opposing perspective of recipients in order to facilitate mutual understanding or agreement.

In the following sections, we separately focus on perspectives of recipients (Section 2.2) and perspectives of senders (Section 2.2.1), and how their individual perspectives relate to communication processes. Then, we focus on interactive dynamics between senders and recipients. We focus on how senders and recipients collaborate to achieve mutual understanding and/or agreement (Section 2.4), and on what it takes to change the perspectives of individuals. Finally, we link our discussion to the Perspective Web (Section 2.5).

2.2 Recipient Perspectives: The Role of Perspectives in Processing Incoming Information

People are exposed to large amounts of information on a daily basis. Out of sheer cognitive necessity, they will not grant all information equal attention. Also, they will not process information to an equal extent. In determining what information to assign cognitive resources to, perspectives play a major role.

In psychological theory, the distinction is made between voluntary and involuntary attention (Kahneman, 1973). When people attend to information voluntarily, they do so because of a conscious choice process. They find the information interesting, for example, or exciting or enjoyable, or they find its source nice or trustworthy; all these assessments can be easily linked to people's perspectives. When people attend to information involuntarily, perspectives might also be important. In some cases, information may grab people's attention simply because it is vivid – that is, perceptually or emotionally arousing – such as colorful or fear-eliciting information. In other cases, involuntary attention is granted to information that is salient – that is, information that we have learned to attend to (Tversky and Kahneman, 1973). For example, we may note a particular car in a parking lot, not because of its color or size, but because it is the same model and color as a car we used to own. Here, personal (or cultural, or social) perspectives come into play. We tend to pay attention to information that we have learned to pay attention to, and for that reason we also ignore a lot of information that we have learned to ignore. As a result, perspectives may determine for a significant part what information we perceive to be there. This is a notion that we should not underestimate: information may be present right before our eyes and we may not see it. Compelling examples are change blindness, where people fail to see changes in stimuli (landscapes, and even persons) after the briefest of intermissions (Simons and Rensink, 2005), and banner blindness, where people perceptually "block" ad banners while surfing the Web (Benway, 1998).

Once information has passed the barrier of our attention, it might still be processed only superficially. The Elaboration Likelihood Model (Petty et al., 1983) posits that people are "cognitive misers" who are quite selective in assigning their cognitive resources to processing particular information. As a result, we employ two different routes of message elaboration – one that uses a lot of cognitive resources and one that uses only a few. In the "central" route, using many resources, we process information rigorously, consider arguments and evidence, and come to a deliberate conclusion. In the "peripheral" route, using few resources, we process information only superficially. We consider peripheral cues (message source, pictures, or communication style) rather than arguments and evidence, but in this way may still make up our mind. Which elaboration route is used may depend on the information (e.g., we tend to process ads more superficially than scientific papers), on the context (e.g., we tend to resort to peripheral processing when we have little time, or if information is too abundant), on personality (e.g., people who are low in need for cognition resort to peripheral processing more often), and on mood (people in a positive mood are more likely to process information peripherally), but above all on our perspectives. If we find the subject matter unimportant, or if we feel we lack the background knowledge to effectively process the information, we will resort to peripheral processing. If we think the message content is important, and we are able to understand it, we will resort to central processing.

Clearly, using the one processing route or the other on the same information may lead to rather different outcomes. Many ads, for example, are quite effective in eliciting purchase intentions when viewers process them peripherally. In that case, cues (e.g., celebrities, nice music, happy images) do their work in getting the message across that the advertised product will have a positive effect on one's quality of life. However, when people watch the same ads while processing the information centrally, the arguments and evidence (if present) may be perceived as shallow and manipulative, and the cues hardly weigh in.

In sum, our perspectives are key in determining which information we attend to, and to which information we assign our cognitive resources. This means that not all information will be attended to and not all arguments will be considered – to some extent, we naturally create a mental "bubble" of information that we spend time on to mull over. That does not mean the other information does not influence us. Peripheral information processing may also lead to cognitive changes, for example, to liking a particular brand or product. And there is even proof that information that we did not consciously perceive may influence our thought process, for example, through "priming" (Strahan et al., 2002). However, it is important to acknowledge that our minds are constituted in a conservative way, in the sense that the perspectives we already hold for a large part determine what we will attend to and what we filter out, what we find worthwhile to mull over deeply or think about only superficially, and thus also whether or not we should change our mind about something.

2.2.1 Perspectives in Information Seekers

So far in this section we have considered recipients as relatively passive "receivers" of information, whose main job challenge was to filter out information from an information-rich environment. However, information recipients may also actively seek information about subjects that they are interested in, or need, for example, health information or product reviews. In this search process, too, recipient perspectives play a major role. Many studies have shown tendencies for information seekers to favor some information over other information. Well-known examples are that people favor arousing (e.g., sensational) over tedious information, personal (e.g., narrative, relatable) over impersonal information, and negative over positive information (Sen and Lerman, 2007). Perhaps more surprisingly, people also favor information that they agree with over information that they do not agree with. This is explained by two processes: selective exposure (Stroud, 2008) and confirmation bias (Nickerson, 1998). Selective exposure describes that people tend to seek out media sources and messages that are in line with their own attitudes and beliefs; that is, sources and messages that are congruent with their own perspectives. For example, Republican voters in the United States tend to gravitate toward Fox News as a main news outlet, whereas Democratic US voters would usually rather watch CNN (Iyengar and Hahn, 2009). In a similar vein, the notion of confirmation bias describes that also within messages, people tend to assign more weight to elements that confirm their opinion than to elements that disconfirm it. Well-known examples can be found in the processing of online reviews; as soon as we decide we like, for example, a particular restaurant, we will tend to focus on descriptions of its positive characteristics (confirming our opinion) and ignore or dismiss descriptions of its negative characteristics (Yin et al., 2016).

Similar to the effects of selective attention and elaboration likelihood described earlier, selective exposure and confirmation bias also cause information seekers to create private information bubbles, in which it becomes relatively unlikely that information that is incongruent with one's own perspective is taken into consideration. Digital information platforms further reinforce this process by offering personalized information feeds. This personalization involves filtering out information that is unlikely to be "clicked on," and thus, in effect, filtering out information that is incongruent with one's own perspective. As a result, for people in search of information on digital platforms, information that confirms their own perspectives will be not only psychologically more salient due to selective exposure and confirmation bias but also physically more salient due to recommendation algorithms.

In sum, recipient perspectives are a major factor in understanding what information is attended to and processed. Information that is not considered personally relevant is easily looked over and will be processed only superficially. Information that is incongruent with existing perspectives is more easily dismissed, ignored, or simply not noticed. The result is that perspectives tend to be self-reinforcing: perspective-congruent information confirms existing ideas, and perspective-incongruent information is unlikely to be seriously considered.

Despite this inherent stability of recipients' perspectives, senders will want to get their message across; if not to persuade, then maybe to inform or simply be heard. The following section discusses how senders incorporate their perspectives in communication, and how senders adjust their message in order to get it across.

2.3 Sender's Perspectives: How a Sender's Perspective is Reflected in Communication and Language Use

Senders can express their perspectives on a given topic in communication in various ways, through verbal or nonverbal means. Senders' messages and language use then reflect their personal point of view and attitude, sometimes without their conscious awareness. Here we will focus on framing and linguistic biases in label use and in behavior descriptions of social category members.

A first way in which people communicate their perspective is by framing (Goffman, 1974). Framing has initially been studied in the context of mass media effects but can play a similar role in interpersonal communication. By highlighting certain events and by placing them within a particular context, a sender can encourage or discourage certain interpretations (Scheufele and Moy, 2000). For instance, when communicating about COVID-19 vaccination, one can choose to focus on the dangers, potential side effects, and hesitations of people to take a vaccine, or focus on the positive consequences of a successful vaccination campaign. Such communicative choices reveal the perspective of the sender, and can thereby influence recipients' interpretations.

Aside from approaching a topic from one or the other broader point of view, framing can also be reflected in more specific communicative choices. Framing has more specifically been defined as "the words, images, phrases, and presentation styles" that senders use when relaying information to recipients (Druckman, 2001). Such communicative choices also follow from a sender's perspective. Research has, for instance, shown how a sender's existing norms as related to the majority–minority context in which one resides are reflected in language use.

Specifically, when communicating about differences between social categories (e.g., men vs. women; gay vs. straight), senders typically focus on the atypical category (e.g., a minority within the relevant domain) in their formulation, which is subsequently compared to the norm category (e.g., a majority). Explanations of differences between gay and straight men, for instance, typically take gay men as the subject, particularly in a context in which straight men are considered the normative majority group (Hegarty and Pratto, 2001). The same has been shown to occur in explanation of gender differences

(Bruckmüller et al., 2012). For instance, when trying to explain a relatively low representation of women in academia, participants focused their explanations on the atypical category (e.g., female professors are …). These studies show that a sender's perspective on who is considered the normative majority and minority group is reflected in subtle linguistic choices. An increased relative focus on minorities in category comparisons can be reflected in language as a combination of placing the atypical category in sentence subject position, mentioning it first in a comparison with a referent (e.g., Bruckmüller and Abele, 2010), or simply in more frequent use of the minority (vs. majority) category label.

Another relevant area of research has focused on how a sender's perspective toward social categories of people is shared with others. A sender's social category stereotypes can be subtly expressed in differences in variations in word choice and formulations (Beukeboom, 2014; Beukeboom and Burgers, 2019). First, a sender's use of category labels reveals how one categorizes the social world, and which social categories are considered as distinct and meaningful "kinds" (Rosch et al., 1976). Labels reveal a sender's level of generalization in which categorization occurs. When talking about immigrants, for instance, one may either use a generic label (immigrants), specify broad subgroups (e.g., non-Western vs. Western immigrants), or more specific subgroups (e.g., second-generation Turkish immigrants), or instead focus on individual cases. Second, the semantic content of a label term reveals a sender's stereotypes and prejudice about a social category. To refer to "immigrants," one can use negative labels like "fortune-seekers," "aliens," "outsiders," or highly negative metaphorical terms like "parasites" (Musolff, 2014). Derogatory labels and social slurs are more likely used when one intends to convey negative prejudice and to qualify categories or category members in a disparaging manner (Maass et al., 2014). The choice of labels thus reveals one's thoughts about social groups.

In addition to the content of label terms used, the linguistic forms of labels reveal a sender perspective on a discussed social category. Experimental research showed a difference in the use and effects of nouns versus adjectives in label use (Carnaghi et al., 2008). Participants who were led to believe that athletic abilities are genetically determined (high essentialism) were more likely to choose a noun label (athlete) than an adjective (athletic) to describe an individual target person active in athletic sports, compared to participants who believed athletic abilities are a transient characteristic and the result of training (low essentialism). Noun labels thus convey that a social category is perceived as a meaningful entity (Beukeboom and Burgers, 2019). Particularly generic noun labels (in plural form) paired with a characteristic (e.g., boys play with trucks; girls are sweet) reveal a sender's generalized stereotype about a category, and can thus communicate it to recipients (Cimpian and Markman, 2009).

Next to labels, the way in which the behaviors of social category members are described can reveal a sender's stereotypic expectancies. Research within this field shows that information about people that is congruent with existing social-category cognition is formulated differently than information that is incongruent with social-category cognition (Beukeboom, 2014; Beukeboom and Burgers, 2019). First, various studies have shown a stereotype consistency bias, showing that stereotype-congruent information tends to be shared more prominently compared to stereotype-incongruent information about categorized individuals (Klein et al., 2008; Schaller et al., 2002). This occurs in interpersonal conversations, but mass media can also be a powerful transmitter of stereotypic characteristics associated with categories (Arendt, 2013; Ramasubramanian, 2011). For instance, television commentators covering professional football and basketball have been shown to subtly induce associations by more frequently emphasizing athleticism of African American players, while emphasizing intellectual abilities and character traits of white players (Rada and Wulfemeyer, 2005). The linguistic form of behavior descriptions can also reveal a sender's stereotypes about described individuals. Research has revealed several subtle variations in verbal formulation that communicate whether a described behavior is expected rather than unexpected for a categorized target (Beukeboom, 2014; Beukeboom and Burgers, 2019).

Research on the linguistic intergroup bias (Maass et al., 1995) and linguistic expectancy bias (Wigboldus et al., 2000) shows that language abstraction varies as a function of stereotype consistency. A target's behavior is more likely described in abstract terms (e.g., adjectives) when it is stereotype-congruent and the fit between the described behavior and existing stereotypic expectancies is high (e.g., the woman is emotional), but more concretely when it is stereotype-incongruent and fit is low (e.g., descriptive action verbs; the man is crying).

The stereotypic explanatory bias relates to a comparable linguistic variation and reveals that senders tend to produce more explanatory comments for stereotype-incongruent (vs. congruent) behaviors (e.g., the man is crying because he has had a rough day; Sekaquaptewa et al., 2003). Such explanations of stereotype-incongruent behavior are aimed at clarifying the apparent inconsistency, which is not needed for stereotype-congruent behavior. Both more concrete and situational, and explanatory language used for stereotype-incongruent behaviors imply that these are just one-time events, and this is also what recipients infer (Beukeboom and Burgers, 2019).

In communication about stereotype-incongruent information, senders can also introduce stereotype-congruent terms by using negations. Research on the negation bias (Beukeboom et al., 2010, 2019) revealed that the use of syntactic negations (e.g., not stupid, rather than smart) is more likely in descriptions of stereotype-inconsistent compared to stereotype-consistent behaviors. For example, if a sender's stereotypic expectancy dictates that drug addicts are deceitful, but a particular addict violates this expectancy by showing very honest behavior, the sender is likely to reveal his prior expectancy by using a negation like "The drug addict was not deceitful." In contrast, for stereotype-congruent behavior (e.g., the drug addict was deceitful; the priest was honest), the use of negations is less likely. Negations thus allow senders to introduce stereotype-congruent concepts in communications about stereotype-incongruent information, and thereby reaffirm existing associations with a category.

In sum, senders' perspectives can be expressed in various ways in communication: both in linguistic content and in linguistic form. In general, it can be said that senders are more likely to share information that is congruent with their existing perspective, and when perspective-incongruent information is shared, one tends to subtly indicate that it concerns unexpected, transient information. Often a sender will not be aware of this, yet it may have an effect on how recipients view the information, and thereby maintain existing perspectives.

2.4 Social and Interactive Dynamics

So far, we separately discussed how perspectives of recipients and senders relate to communication. However, when recipients and senders rely on their personal perspectives, they are usually situated in a social context. In social situations, senders adapt and tune their utterances and communication to an intended or perceived recipient or broader audience. Likewise, the recipient will adapt their interpretation to what they assume to be the sender's perspective and intention. Moreover, depending on the medium, the recipient's feedback can influence the sender's message. Our communicative messages, and the interpretation of these messages, are thus often not just a direct expression of personally held attitudes and beliefs. Instead, they are often the result of an adaptation to an (assumed) perspective of an interaction partner.

In communication, the goal is often to align our perspectives with each other, and thereby achieve mutual understanding and/or agreement about the topic. Mutual understanding means that sender and recipient share the same knowledge (i.e., beliefs) on a given topic; they have reached common ground (Clark and Brennan, 1991). Agreement is about aligning attitudes and opinions about a topic and should be distinguished from mutual understanding. In a discussion about vaccination, for instance, two persons may have the same knowledge available about the topic, they may also be completely knowledgeable about each other's beliefs on which they form their opinions (i.e., mutual understanding), yet they can still completely disagree. That is, one person could have a negative opinion about getting vaccinated, while the other person is positive.

2.4.1 Adapting Communication Content to Recipient Perspectives to Achieve Mutual Understanding and Agreement

When we communicate with others, we need to take the perspective of the people we are communicating with. This can be a direct conversation partner when using an interactive medium, but even when we write a website text or mass email, we often have a picture of our audience in mind. Perspective taking is one of the most important requirements for successful communication (Holtgraves, 2002). To get a message across, a sender must consider whether a recipient can understand what they are saying. At the same time, a recipient must consider the perspective of the sender to understand the intention behind an utterance (what could they mean given their perspective?).

How does this work? An important area of research has revealed that conversation partners do this by constantly trying to determine what is common ground (Clark, 1996). Common ground refers to the knowledge that conversation partners share – the information that conversation partners mutually know, believe, and recognize. It has been shown that people rely on heuristics or rules of thumb that allow them to assess what is part of their common ground (Clark and Brennan, 1991). One of these rules of thumb that people rely on is the heuristic of physical copresence. If two people are physically together and attending to roughly the same aspect of their environment, they assume that what both perceive is common ground. When someone is passing by in front of you, you can assume that the person's behavior and appearance are mutually known, and you consequently do not need to describe these to your conversation partner. A second rule of thumb is the so-called heuristic of linguistic copresence. This means that people generally assume that information previously introduced during a conversation is part of the common ground. This simply means that once you have mentioned something, you do not need to explain it again.

A third finding is that people draw immediate assumptions about common ground from their conversation partner's community membership. The categorization of a person in a certain social category or community (e.g., nationality, age, gender, occupation) activates a stereotype that allows one to make reasonable assumptions about what the other person knows or believes. When strangers meet, they will immediately try to categorize each other and make generic assumptions about the other person's knowledge. Based on each other's appearances (a police officer, a man in a white coat, a little kid) or the situation in which they happen to be (behind a post office desk, in a soccer stadium), people make immediate inferences. Based on these inferences, people automatically adapt their utterances. Without contemplation you know you can ask a doctor about health issues, and two English speakers can assume they are both familiar with the meaning of English vocabulary. Even on the basis of a foreign dialect alone (e.g., Kingsbury, 1968), people draw immediate inferences about what their conversation partner should know, and formulate a message that is understandable in light of such knowledge.

An example of such adaptations in communication is that with accumulating common ground, people change from indefinite descriptions, like "a woman" or "some shops," to definite descriptions, like "the woman" or "those shops" (Linde and Labov, 1975). The first mention of something, when it is not yet part of the common ground, is usually done with indefinite articles (a, an). Later, when a topic has been grounded (i.e., added to the common ground; Clark and Brennan, 1991), definite descriptions are used. This can also be seen in how stories are built up: "This is a story about a girl. The girl lived in a big castle."

Similarly, in interactive conversations, when speakers are trying to explain what they are talking about, they first tend to offer a conceptualization that is tentative or provisional, using hedges such as sort of, kinda, looks like (e.g., "the guy that acted kinda like Mr. Bean, you know?"). Once conversation partners agree to a conceptualization, they drop the hedges and use more definite descriptions (e.g., the Bean guy; Brennan and Clark, 1996).

Related to this is another important finding, namely that with accumulating common ground conversations gain in efficiency. Conversation partners need fewer words, fewer turns, and less time to get mutual understanding of a topic. You will not need to tell your husband the whole story about your mutual holiday to France in great detail because you share the same perspective. When referring to this mutual experience you can simply say "just like France," and he will understand. In other words, when common ground between conversation partners increases, they can understand each other's perspective more easily (Schober and Brennan, 2003).

The theory on common ground also helps one to understand how misunderstanding and miscommunication arises, as these are often the result of a failure to properly consider common ground. One factor is that relying on general rules of thumb or heuristics in assessing common ground can lead to errors. First, the reliance on the heuristics of physical and linguistic copresence does not guarantee a correct judgment. A conversation partner may have missed the person passing by and have forgotten something you previously mentioned. Likewise, the social categorization of a conversation partner will not always lead to a correct inference as generic stereotypic expectancies do not always fit particular individuals. We have beliefs about what elderly people are like in general, but this does not mean that each elderly individual is exactly like that. As a result, people may provide a very detailed explanation about computer use to an elderly woman although she may be an IT specialist. Similarly, if someone speaks with a particular foreign accent, this may not be related to their knowledge of or opinion on a particular topic.

Other types of miscommunication arise from a type of egocentrism in which the other person's perspective is not sufficiently taken into account. According to Keysar (2007), the default tendency is to speak and listen from an egocentric perspective. Taking another person's perspective requires ability and motivation that people often do not have (Horton and Keysar, 1996). Yet, depending on the communication medium, conversation partners can immediately repair and correct misunderstandings once they become clear. In synchronous conversations, we provide each other with feedback about whether something is understood or not, and thereby trigger the other to better adapt to one's perspective by providing more detail or additional information. Feedback, using vocal (e.g., uhuh, yeah, ok) and nonverbal signals (e.g., nodding, a smile or frown), is crucial for having a smooth conversation (Clark and Brennan, 1991). This process in which listeners inform speakers about their level of understanding usually occurs automatically and outside awareness. Yet, these subtle (nonverbal) reactions are very important, and even in situations in which one person is talking and the other is merely listening, speakers constantly monitor their conversation partner and change their story depending on the listeners' replies and feedback (Bavelas et al., 2000; Beukeboom, 2009).

Communication partners adapt their communication in order to not only achieve mutual understanding with each other but also to coordinate agreement. Research in line with Politeness theory (Brown and Levinson, 1987) shows that people often have a strong preference for agreement with their communication partners, and consequently tend to avoid or mitigate disagreement in various ways. A sender may, for instance, mitigate an expressed opinion when communicating with someone who apparently has an opposing perspective on the topic. A motivation to reach consensus and agreement with a conversation partner may follow from communicators' wish to form or affirm a social relationship and to show in-group solidarity (Clark and Kashima, 2007). When these social goals prevail, a formulated message may not even reflect the sender's own attitudes and beliefs about a given topic, but instead reflect the assumed perspective of the audience or community. Similarly, the social norms that prevail in a community may determine the appropriate communication content. For instance, the choice to refer to social groups using derogatory labels depends on whether it is considered appropriate in the communicative context (Crandall et al., 2002; Fasoli et al., 2015) and whether senders expect recipients to endorse such expressions.

In sum, maintaining mutual understanding and/or agreement is a dynamic collaborative process in which communication partners attempt to take each other's perspective when they can. When doing so they basically attempt to build a common perspective, which is needed to have a smooth and pleasant conversation. How they can achieve this depends on the medium. Only interactive media allow recipients to provide feedback to allow a sender to immediately adapt one's messages to the recipient's perspective.

2.4.2 Changing Perspectives

In prior sections we discussed the inherent stability of recipients' perspectives by focusing on attention and information processing: perspectives tend to reinforce themselves by affecting the selection and processing of incoming information. We also discussed that senders are more likely to share and maintain existing perspectives than to introduce perspective-incongruent information. The interactive dynamics we discussed also suggested that people often tend to seek mutual understanding and agreement, rather than confronting each other with opposing perspectives, in order to have pleasant interactions. All these processes work toward the maintenance of existing perspectives.

Nevertheless, in order for societies to advance, it is necessary that senders sometimes succeed in changing recipients' minds. To move forward, people sometimes have to adjust their perspectives, and try new things. In this section, we discuss how perspectives can be changed. We start at the level of basic cognitive elements – beliefs and attitudes – and subsequently discuss the potential to change broader perspectives.

2.4.3 Theories about Attitude Change

In the early days of communication science and social psychology, many researchers were adamant to understand how, for example in Nazi Germany, entire populations seemed so easily swayed into supporting very extreme political leaders and ideas. The basic paradigm in mass communication at that time seemed to be that audiences were very easily influenced and therefore mass media had a great responsibility to be a force of good – for example, progress, education, cooperation – rather than bad. Later empirical researchers, however, found out that audiences were actually quite hard to influence, even when you had their entire attention and processing resources at your disposal. These anomalous findings led to a shift from mass communication as a "hypodermic needle," to the "minimal effects" paradigm. The widely acknowledged idea about communication effects became that senders' messages did not so much affect recipients' ideas, but rather that recipients' ideas affected how they responded to senders' messages.

This generic idea was also strongly anchored in persuasion theories from the 1950s and 1960s by seminal authors such as Festinger (1957), studying the human tendency to rationalize away dissonant cognitions, Hovlandt, who looked at the impact of perceived source credibility, and Hovland and Weiss (1951) and Greenwald (1968), who argued that we tend to perceive our cognitive response to a message rather than the content of the message itself. All these theories are still very much accepted today.

Another central persuasion theory fitting into the minimal effect paradigm is Sherif and Sherif's social judgment theory (SJT) (Sherif and Sherif, 1967). SJT posits that "the basic information for predicting a person's reaction to a communication is where he places its position and communicator relative to himself." In attempting to predict to what extent recipients of messages will respond to messages with an attitude change, SJT introduces the idea of attitudes as (two-dimensional) structures rather than as (one-dimensional) points. That is, whereas most attitude models conceptualize an attitude as a point on an axis (e.g., an evaluation of an object or a degree of agreement with a statement), Sherif and Sherif conceptualize attitudes as a series of latitudes around this "anchor" point. Recipient responses to a message, subsequently, depend on the latitude in which the message is categorized.

There are basically three possibilities. First, the message could fall into the latitude of acceptance, which has the recipient's anchor as its center, and which constitutes all evaluations that are perceived as sufficiently congruent to the anchor point. If message content is categorized in this latitude, an assimilation effect takes place. The message is perceived as consistent to the recipient's attitude, and no response in terms of an attitude shift will occur. At the extremes of the axis, there are latitudes of rejection, constituting all evaluations that are perceived as incongruent to the anchor of the recipient. If message content is categorized in this latitude, a contrast effect takes place. Again, no shift in attitude is expected, but this time because the message is perceived as entirely inconsistent to the recipient's anchor. Different from assimilation effects, contrast effects involve negative affective responses as well; to the message and – if identifiable – its source. Finally, there are (usually two) latitudes of noncommitment. If message content is categorized here, the recipient does not immediately respond but would need to make up his/her mind before responding. SFI posits that only messages that fall into the latitude of noncommitment may potentially change recipients' beliefs and attitudes, although not necessarily so. Thus, the original attitude of a recipient determines to a large part the response to the communication; only messages that fall into the recipient's "sweet spot," the latitude of noncommitment, may be effective in establishing an attitude change. Finding this sweet spot is of course not impossible. For many senders aiming to be persuasive (car salesmen, politicians, and parents alike), it is part of the interactive dynamics to find this spot in recipients' attitude structures and deliver exactly the message that is most effective in changing the recipients' attitudes. This is another reason why senders often align their message with the recipient's perspective rather than with their own.

2.4.4 Changing Broader Perspectives

There is not much theory in psychology or communication on how recipients' broader perspectives can potentially be changed. Based on knowledge about larger attitude structures or networks, it is, however, easy to predict that these are likely much more stable than specific individual attitudes. Mechanisms such as cognitive dissonance and balance theory (Heider, 1959) play a role here, as discussed in Section 2.2. The denser and more intertwined our attitude networks are, the more beliefs and attitudes will be linked, and the stronger the pull toward consistency will be. Swaying people from, say, progressive to conservative or from climate skeptic to environmentalist in a short period of time, therefore, will hardly be possible. A possible exception seems to be the recent development where people, in an apparently short time frame, come to support strong anti-establishment and conspiracy-related ideas (e.g., QAnon). Supporters of such ideas often describe themselves as being "red-pilled" – they experienced a sudden disruption in their perspective and now have a different view on a great number of issues. Current psychological and communication theory provides no theoretical mechanism that would explain such large persuasive effects; literature on indoctrination and brainwashing describes very specific psychological procedures rather than psychological mechanisms (Baron, 2000). Clearly, it is an important task for scholars to fill this theoretical gap.

In sum, most basic persuasion theory, including classic theories such as cognitive dissonance, source credibility theory, cognitive response theory, and SJT, all predict that perspectives, as present in the minds of recipients of communication, have a strong influence on how incoming information is perceived, and on whether this information is likely to have any influence. SJT makes clear that only information that is neither too alien nor too familiar to recipients' own ideas has a potential to be influential. Changing even recipients' most basic perspectives (individual attitudes and beliefs) with information – even if attended to and processed – therefore is quite a challenge for senders. However, this challenge can be met by senders who are able to adjust their message to the attitude structure of the recipient. More broad and abstract perspectives may be even harder to sway because they consist of extensive mental networks of beliefs and attitudes that tend to naturally gravitate toward stability.

2.5 Conclusion

In this chapter, we discussed perspectives as exchanged by senders and recipients through messages. We conceptualized perspectives as beliefs and attitudes, or networks thereof, which could be specific and concrete or broad and abstract. We argued that perspectives that we hold as human beings are relatively stable because we have a tendency to favor perspective-congruent information over perspective-incongruent information and are quite insensitive to opinions that differ from ours. Broader and more abstract embedded perspectives are even less likely to change. Nevertheless, it is possible for senders and recipients to interact. Senders disseminate perspectives in their communication, for instance, by (often unconsciously) using particular labels and linguistic constructions. Also, communication partners have several techniques at their disposal to find a shared perspective, which increases mutual understanding and facilitates an exchange of ideas.

2.5.1 Personal Perspectives and the Perspective Web

What would the development of Perspective Web technology mean for its users? Most certainly, the Perspective Web would benefit communication scholars interested in the exchange, evolution, and cross-fertilization of perspectives through the social process of communication. With a Perspective Web, we could trace back perspectives to an original source or event, relate or contrast perspectives, and build networks of associated attitudes and beliefs that are potentially as intricate as people may hold them in their minds. We could make the perspectives that are implicitly communicated through language use explicit, we could analyze perspectives of people in opposing groups and see where they may find common ground, and we may even come to understand how people, despite their naturally stable perspectives, are sometimes radically swayed.

For nonscientist users, the Perspective Web may help to assess particular information they encounter online or clarify its sender's perspective and perhaps even intentions. The Perspective Web may even help users to better understand themselves. It could, for example, make explicit that one's language harbors implicit stereotypes. In many cases, people are unaware of any stereotypes they may hold, but once these are made explicit, they are highly motivated to reduce them. Also, through a Perspective Web, people may be made aware of how their use of language comes across to others. Furthermore, the Perspective Web may help people to find common ground and to find common understanding, and perhaps assist people in attempts to persuade others to adopt new and innovative ideas.

More controversial societal effects may also be expected: Perspective Web technology may be very beneficial to companies that have exclusive access to large amounts of user communication data, for example, social media platforms and other so-called Big Tech companies. These companies can conduct similar analyses as scientists, but then outside of the public space and eye, and with an entirely different motivation. In fact, understanding individual people's perspectives deeply and on a massive scale may be the holy grail to increase the effectiveness of content recommendation algorithms, further personalized advertising, and exerting political influence. Needless to say that these developments will inflate negative effects currently attributed to the sector: excessive social media use, filter bubbles, commercialization of social interactions, and an undesirable control over information flows. These dystopian future visions are not unique to Perspective Web technology – however, its immense potential strongly illustrates the importance for users to regain control over all of their own data.

In sum, the Perspective Web merges ideas from computational linguistics, human psychology, language, and communication into a powerful vision for technological solutions that lead to better understanding of, and between, people.

References

Arendt, F. 2013. Dose-dependent media priming effects of stereotypic newspaper articles on implicit and explicit stereotypes. *Journal of Communication*, **63**(5), 830–851.

Baron, R. S. 2000. Arousal, capacity, and intense indoctrination. *Personality and Social Psychology Review*, **4**(3), 238–254.

Bavelas, J. B., Coates, L., and Johnson, T. 2000. Listeners as co-narrators. *Journal of Personality and Social Psychology*, **79**, 941–952.

Benway, J. P. 1998. Banner blindness: The irony of attention grabbing on the World Wide Web. Pages 463–467 of: *Proceedings of the Human Factors and Ergonomics Society Annual Meeting*, Vol. 42, No. 5. Los Angeles, CA: Sage.

Beukeboom, C. J. 2009. When words feel right: How affective expressions of listeners change a speaker's language use. *European Journal of Social Psychology*, **39**(5), 747–756.

Beukeboom, C. J. 2014. Mechanisms of linguistic bias: How words reflect and maintain stereotypic expectancies. Pages 313–330 of: Forgas, J. P., Vincze, O., and László, J. (eds), *Social cognition and communication*. New York, NY: Psychology Press.

Beukeboom, C. J. and Burgers, C. 2019. How stereotypes are shared through language: A review and introduction of the social categories and stereotypes communication (SCSC) framework. *Review of Communication Research*, **7**, 1–37.

Beukeboom, C. J., Finkenauer, C., and Wigboldus, D. H. J. 2010. The negation bias: When negations signal stereotypic expectancies. *Journal of Personality and Social Psychology*, **99**(6), 978–992.

Beukeboom, C. J., Burgers, C., Szabó, Z. P., Cvejic, S., Lönnqvist, J.-E. M., and Welbers, K. 2019. The negation bias in stereotype maintenance: A replication in five languages. *Journal of Language and Social Psychology*, **39**(2), 219–236.

Brennan, S. E. and Clark, H. H. 1996. Conceptual pacts and lexical choice in conversation. *Journal of Experimental Psychology: Learning, Memory, and Cognition*, **22**, 1482–1493.

Brown, P. and Levinson, S. C. 1987. *Politeness: Some universals in language usage* (pp. xiv, 345). Cambridge: Cambridge University Press.

Bruckmüller, S. and Abele, A. E. 2010. Comparison focus in intergroup comparisons: Who we compare to whom influences who we see as powerful and agentic. *Personality and Social Psychology Bulletin*, **36**, 1424–1435.

Bruckmüller, S., Hegarty, P., and Abele, A. E. 2012. Framing gender differences: Linguistic normativity affects perceptions of power and gender stereotypes. *European Journal of Social Psychology*, **42**(2), 210–218.

Carnaghi, A., Maass, A., Gresta, S., Bianchi, M., Cadinu, M., and Arcuri, L. 2008. Nomina sunt omina: On the inductive potential of nouns and adjectives in person perception. *Journal of Personality and Social Psychology*, **94**(5), 839–859.

Cimpian, A. and Markman, E. M. 2009. Information learned from generic language becomes central to children's biological concepts: Evidence from their open-ended explanations. *Cognition*, **113**(1), 14–25.

Clark, A. E. and Kashima, Y. 2007. Stereotypes help people connect with others in the community: A situated functional analysis of the stereotype consistency bias in communication. *Journal of Personality and Social Psychology*, **93**, 1028–1039.

Clark, H. H. 1996. *Using language*. Cambridge: Cambridge University Press.

Clark, H. H. and Brennan, S. E. 1991. Grounding in communication. Pages 127–149 of: Resnick, B., Levine, J., and Behrend, S. D. (eds), *Perspectives on socially shared cognition*. Washington, DC: American Psychological Association.

Crandall, C. S., Eshleman, A., and O'Brien, L. 2002. Social norms and the expression and suppression of prejudice: The struggle for internalization. *Journal of Personality and Social Psychology*, **82**, 359–378.

Dalege, J., Borsboom, D., van Harreveld, F., van den Berg, H., Conner, M., and van der Maas, H. L. 2016. Toward a formalized account of attitudes: The causal attitude network (CAN) model. *Psychological Review*, **123**(1), 2–22.

Druckman, J. N. 2001. The implications of framing effects for citizen competence. *Political Behavior*, **23**(3), 225–256.

Eagly, A. H. and Chaiken, S. 1995. Attitude strength, attitude structure, and resistance to change. *Attitude strength: Antecedents and consequences*, **4**(2), 413–432.

Fasoli, F., Carnaghi, A., and Paladino, M. P. 2015. Social acceptability of sexist derogatory and sexist objectifying slurs across contexts. *Language Sciences*, **52**, 98–107.

Fazio, R. H. 1995. Attitudes as object-evaluation associations: Determinants, consequences, and correlates of attitude accessibility. Pages 247–282 of: Petty, R. E., and Krosnick, J. A. (eds), *Attitude strength: Antecedents and consequences, Vol. 4.* Lawrence Erlbaum Associates, Inc.

Festinger, L. 1957. *A theory of cognitive dissonance, Vol. 2*. Stanford, CA: Stanford University Press.

Goffman, E. 1974. *Frame analysis: An essay on the organization of experience*. Cambridge: Harvard University Press.

Greenwald, A. G. 1968. Cognitive learning, cognitive response to persuasion, and attitude change. Pages 147–170 of: Brock, T. C. and Ostrom, T. M. (eds.), *Psychological foundations of attitudes*. New York and London: Academic Press Inc.

Haugtvedt, C. P. and Petty, R. E. 1992. Personality and persuasion: Need for cognition moderates the persistence and resistance of attitude changes. *Journal of Personality and Social Psychology*, **63**(2), 308.

Hegarty, P. and Pratto, F. 2001. The effects of social category norms and stereotypes on explanations for intergroup differences. *Journal of Personality and Social Psychology*, **80**, 723–735.

Heider, F. 1959. On perception, event structure, and the psychological environment. *Psychological Issues*, **1**(3), 1–123.

Holtgraves, T. M. 2002. *Language as social action: Social psychology and language use*. Mahwah, NJ: Erlbaum.

Horton, W. S. and Keysar, B. 1996. When do speakers take into account common ground? *Cognition*, **59**, 91–117.

Hovland, C. I. and Weiss, W. 1951. The influence of source credibility on communication effectiveness. *Public Opinion Quarterly*, **15**(4), 635–650.

Iyengar, S., and Hahn, K. S. 2009. Red media, blue media: Evidence of ideological selectivity in media use. *Journal of Communication*, **59**(1), 19–39.

Kahneman, D. 1973. *Attention and effort*. Englewood Cliffs, NJ: Prentice-Hall.

Keysar, B. 2007. Communication and miscommunication: The role of egocentric processes. *Intercultural Pragmatics*, **4**(1), 71–84.

Kingsbury, D. 1968. Manipulating the amount of information obtained from a person giving directions. Ph.D. thesis, Unpublished Honors Thesis, Department of Social Relations, Harvard University.

Klein, O., Tindale, S., and Brauer, M. 2008. The consensualization of stereotypes in small groups. Pages of 263–292: Kashima, K., Fiedler, K., and Freytag, P. (eds), *Stereotype dynamics: Language-based approaches to the formation, maintenance, and transformation of stereotypes.* Hillsdale, NJ: Lawrence Erlbaum Associates.

Linde, C. and Labov, W. 1975. Spatial networks as a site for the study of language and thought. *Language*, **51**, 924–939.

Maass, A., Milesi, A., Zabbini, S., and Stahlberg, D. 1995. Linguistic intergroup bias: differential expectancies or in-group protection? *Journal of Personality and Social Psychology*, **68**, 116.

Maass, A., Suitner C., and Merkel, E. 2014. Does political correctness make (social) sense? Pages 331–346 of: Laszlo, J., Forgas, J. P., and Vincze, O. (eds), *Social cognition and communication*. New York, NY: Psychology Press.

Monroe, B. M. and Read, S. J. 2008. A general connectionist model of attitude structure and change: The ACS (Attitudes as Constraint Satisfaction) model. *Psychological Review*, **115**(3), 733.

Musolff, A. 2014. Metaphorical parasites and "parasitic" metaphors: Semantic exchanges between political and scientific vocabularies. *Journal of Language and Politics*, **13**, 218–233.

Nickerson, R. S. 1998. Confirmation bias: A ubiquitous phenomenon in many guises. *Review of General Psychology*, **2**(2), 175–220.

Olson, M. A. and Fazio, R. H. 2001. Implicit attitude formation through classical conditioning. *Psychological Science*, **12**(5), 413–417.

Petty, R. E. and Briñol, P. 2015. Emotion and persuasion: Cognitive and meta-cognitive processes impact attitudes. *Cognition and Emotion*, **29**(1), 1–26.

Petty, R. E., Cacioppo, J. T., and Schumann, D. 1983. Central and peripheral routes to advertising effectiveness: The moderating role of involvement. *Journal of Consumer Research*, **10**(2), 135–146.

Pomerantz, E. M., Chaiken, S., and Tordesillas, R. S. 1995. Attitude strength and resistance processes. *Journal of Personality and Social Psychology*, **69**(3), 408.

Rada, J. A. and Wulfemeyer, K. T. 2005. Color coded: Racial descriptors in television coverage of intercollegiate sports. *Journal of Broadcasting & Electronic Media*, **49**, 65–85.

Ramasubramanian, S. 2011. The impact of stereotypical versus counterstereotypical media exemplars on racial attitudes, causal attributions, and support for affirmative action. *Communication Research*, **38**, 497–516.

Rosch, E., Mervis, C. B., Gray, W. D., Johnson, D. M., and Boyes-Braem, P. 1976. Basic objects in natural categories. *Cognitive Psychology*, **8**, 382–439.

Schaller, M., Conway, III, L. G. and Tanchuk, T. L. 2002. Selective pressures on the once and future contents of ethnic stereotypes: Effects of the communicability of traits. *Journal of Personality and Social Psychology*, **82**, 861–877.

Scheufele, D. A. 2000. Agenda-setting, priming, and framing revisited: Another look at cognitive effects of political communication. *Mass Communication & Society*, **3**(2), 297–316.

Schober, M. F. and Brennan, S. E. 2003. Processes of interactive spoken discourse: The role of the partner. Pages 123–164 of: Graesser, A. C., Gernsbacher, M. A., and Goldman, S. R. (eds), *Handbook of discourse processes*. Hillsdale, NJ: Lawrence Erlbaum.

Schwarz, N. 2007. Attitude construction: Evaluation in context. *Social Cognition*, **25**(5), 638–656.

Sekaquaptewa, D., Espinoza, P., Thompson, M., Vargas, P., and von Hippel, W. 2003. Stereotypic explanatory bias: Implicit stereotyping as a predictor of discrimination. *Journal of Experimental Social Psychology*, **39**, 75–82.

Sen, S. and Lerman, D. 2007. Why are you telling me this? An examination into negative consumer reviews on the Web. *Journal of Interactive Marketing*, **21**(4), 76–94.

Sherif, M. and Sherif, C. W. 1967. Attitude as the individual's own categories: The social judgment-involvement approach to attitude and attitude change. Pages 105–139 of: C. W. Sherif (ed.), *Attitude, Ego-Involvement, and Change*. New York: Wiley.

Shultz, T. R. and Lepper, M. R. 1996. Cognitive dissonance reduction as constraint satisfaction. *Psychological Review*, **103**(2), 219.

Simons, Daniel J. and Rensink, R. A. 2005. Change blindness: Past, present, and future. *Trends in Cognitive Sciences*, **9**(1), 16–20.

Smith, E. R. and Semin, G. R. 2004. Socially situated cognition: Cognition in its social context. Pages 53–117 of: Zanna, M. P. (ed), *Advances in experimental social psychology, Vol.* 36. Amsterdam: Elsevier Academic Press.

Strahan, E. J., Spencer, S. J., and Zanna, M. P. 2002. Subliminal priming and persuasion: Striking while the iron is hot. *Journal of Experimental Social Psychology*, **38**(6), 556–568.

Stroud, N. J. 2008. Media use and political predispositions: Revisiting the concept of selective exposure. *Political Behavior*, **30**(3), 341–366.

Tversky, A. and Kahneman, D. 1973. Availability: A heuristic for judging frequency and probability. *Cognitive Psychology*, **5**(2), 207–232.

Wigboldus, D. H., Semin, G. R., and Spears, R. 2000. How do we communicate stereotypes? Linguistic bases and inferential consequences. *Journal of Personality and Social Psychology*, **78**, 5.

Yin, D., Mitra, S., and Zhang, H. 2016. Research note – When do consumers value positive vs. negative reviews? An empirical investigation of confirmation bias in online word of mouth. *Information Systems Research*, **27**(1), 131–144.

3

Computational Linguistics for Subjectivity

Preslav Nakov

3.1 Sentiment Analysis

3.1.1 Background

Internet and the proliferation of smart mobile devices have changed the way information is created, shared, and spreads, for example, microblogs such as Twitter, weblogs such as LiveJournal, social networks such as Facebook, and instant messengers such as Skype and WhatsApp are now commonly used to share thoughts and opinions about anything in the surrounding world. This has resulted in the proliferation of social media content, thus creating new opportunities to study public opinion at a scale that was never possible before.

Do people like the new iPhone? Are there features they complain about? Do Americans support the COVID-19 lockdown measures? What do Republican voters in the USA like/hate about Donald Trump? Do Germans like how Angela Merkel handled the migrant crisis in Europe? How did the Scottish feel about Brexit? Answering such questions requires studying and aggregating the sentiment and the polarity of the opinions people express in social media, which is the topic of study of the well-established and growing field of sentiment analysis.

Sentiment analysis has applications in political science (Borge-Holthoefer et al., 2015; Kaya et al., 2013; Marchetti-Bowick and Chambers, 2012), economics (Bollen et al., 2011; O'Connor et al., 2010), social science (Dodds et al., 2011), and market research (Burton and Soboleva, 2011; Qureshi et al., 2013). It is used to study company reputation online Qureshi et al. (2013), to measure customer satisfaction, to identify detractors and promoters, to forecast market growth (Bollen et al., 2011), to predict the future income from newly released movies, to forecast the outcome of upcoming elections

(Marchetti-Bowick and Chambers, 2012; O'Connor et al., 2010), and to study political polarization (Borge-Holthoefer et al., 2015; Tumasjan et al., 2010).

Before social media, research on opinion mining and sentiment analysis had focused primarily on learning about the language of sentiment in general, meaning that it was either genre-agnostic (Baccianella et al., 2010) or focused on newswire texts (Wiebe et al., 2005) and customer reviews (e.g., from Web forums), most notably about movies (Pang et al., 2002) and restaurants (Pontiki et al., 2014) but also about hotels, digital cameras, cell phones, MP3 and DVD players (Hu and Liu, 2004), laptops (Pontiki et al., 2014), and so on. Nowadays, the focus is primarily on social media, with Twitter being especially popular for research due to its scale, representativeness, variety of topics discussed, as well as ease of public access to its messages (Java et al., 2007; Kwak et al., 2010).

3.1.2 Sentiment Analysis Tasks

There are different formulations of the task that look at different sizes of the target (e.g., words vs. phrases vs. tweets vs. sets of tweets), at different types of semantic targets (e.g., aspect vs. topic vs. overall), at the explicitness of the target (e.g., sentiment vs. stance), at the scale of the expected label (2-point vs. 3-point vs. ordinal), and so on. All these were explored at SemEval, the International Workshop on Semantic Evaluation, which has created a number of benchmark datasets and has enabled direct comparison between different systems and approaches.

Message-Level Sentiment The classic formulation of sentiment analysis is to determine the overall sentiment expressed by a piece of text. Typically, this means choosing one of the following classes to describe the sentiment: POSITIVE, NEGATIVE, NEUTRAL, and OBJECTIVE. Here are some examples:

(1) POSITIVE: *@nokia lumia620 cute and small and pocket-size, and available in the bright test colors of day! #lumiacaption*
(2) NEGATIVE: *I hate tweeting on my iPhone 5 it's so small :(*
(3) NEUTRAL: *The breakfast was sort of OK.*
(4) OBJECTIVE: *The novel is 250 pages long.*

Note that examples (1)–(3) express a subjective sentiment, while example (4) is objective and expresses no sentiment at all. The OBJECTIVE class denotes the fact that the text does not carry any sentiment, for example, example (4) expresses neither a positive nor a negative evaluation of the novel. In contrast, the NEUTRAL class denotes the fact that the textual item does express sentiment and that this sentiment is intermediate between POSITIVE and NEGATIVE, as in example (3), which expresses a lukewarm evaluation about the breakfast served in the hotel. The distinction between subjective and objective language was very clear in early work on sentiment analysis, where sometimes the task was addressed in two steps: (*i*) distinguish between objective vs. subjective language, and (*ii*) predict the polarity for the subjective pieces of text only (Pang and Lee, 2004; Wiebe et al., 2005). However, most subsequent works merged the NEUTRAL and the OBJECTIVE classes due to difficulties for crowdsourcing annotators to see the distinction or simply due to the irrelevance of such a distinction in practical applications.

Next, consider example (5), which makes a joke; such examples are not particularly challenging. Example (6) is also funny, but it is different as it expresses sarcasm and actually conveys the opposite sentiment of what the words it contains suggest. Sarcastic text is particularly challenging for sentiment analysis.

(5) OBJECTIVE: *If you work as a security in a Samsung store... Does that make you guardian of the galaxy??*
(6) NEGATIVE: *I just love missing my train every single day. Really boosts my morale.*

The sentiment of *sarcastic* and *figurative language* tweets was studied in SemEval-2014 Task 9 (Rosenthal et al., 2014) and SemEval-2015 task 11 (Ghosh et al., 2015), and irony detection in SemEval-2018 Task 3 (Van Hee et al., 2018). Sentiment analysis at the message level was studied in a series of tasks at SemEval in 2013–2017 (Nakov et al., 2013, 2016a,b; Rosenthal et al., 2014, 2015, 2017). Classic readings and surveys on sentiment analysis include Kouloumpis et al. (2011); Liu and Zhang (2012); Pang et al. (2002); Pang and Lee (2008); Wiebe et al. (2004).

Sentiment with Respect to a Topic A closer look at examples (1) and (2) shows that the positive/negative sentiment is actually with respect to a topic, *lumia620* and *iPhone 5*, respectively. A topic can be a product (e.g., *iPhone6*), a political candidate (e.g., *Donald Trump*), a policy (e.g., *Obamacare*), an event (e.g., *Brexit*), and so on. Even though tweets are short, they are still long enough to allow the tweet's author to mention several topics and to express potentially different sentiments toward them. For example, in (7) the author is positive about the topic *Donald Trump* but negative about *Hillary Clinton*.

(7) *As a Democrat I couldn't ethically support Hillary no matter who was running against her. Just so glad that its Trump, just love the guy!*
(topic: *Hillary* → NEGATIVE)
(topic: *Trump* → POSITIVE)

Classic reading in topic-based sentiment analysis includes Stoyanov and Cardie (2008). The task was featured as part of SemEval-2015 task 10 (Rosenthal et al., 2015).

Aspect-Based Sentiment Analysis After an even closer look at examples (1) and (2), we can argue that the sentiment is not even about the phone (*lumia620* and *iPhone 5*, respectively), but rather about some specific aspect thereof, namely, SIZE. Similarly, in (8) we can see sentiment with respect to two aspects of the topic *lasagna*: QUALITY (POSITIVE sentiment) and QUANTITY (NEGATIVE sentiment).

(8) *The lasagna is delicious but do not come here on an empty stomach.*

Classic reading on aspect-based sentiment analysis includes Thet et al. (2010). The task was featured at SemEval in 2014–2016 (Pontiki et al., 2014, 2015, 2016).

Learning Sentiment Polarity Lexicons A quick and dirty way to estimate the overall sentiment of a tweet is based on the presence of sentiment-bearing words and phrases, as well as emoticons such as `;)` and `:(`. For this purpose, researchers have been building and using specialized sentiment analysis lexicons, which proved to be quite useful as features. For example, *cute* is a positive word, while *hate* is a negative one, and the occurrence of these words in (1) and (2) can help determine the overall polarity of the respective tweet. Of course, this can be problematic in many cases, for example, when sarcasm is involved as in example (6), or when there are multiple targets as in example (7).

Moreover, many sentiment-bearing words are not universally good or universally bad. For example, the polarity of an adjective could depend on the noun it modifies, for example, *hot coffee* and *unpredictable story* express positive sentiment, while *hot beer* and *unpredictable steering* are negative. Even for the same target noun, the polarity of the modifying adjective could be different in different contexts, for example, *small* is positive in (1) but negative in (2), even though they both refer to a phone.

Despite these limitations, sentiment lexicons have proven to be crucial for opinion mining tasks, especially in early work, when the training datasets were small. Initially, manually crafted sentiment polarity lexicons of small to moderate size were used, for example, *LIWC* (Pennebaker et al., 2001) has 2,300 words, the *General Inquirer* (Stone et al., 1966) contains 4,206 words, *Bing Liu's lexicon* (Hu and Liu, 2004) includes 6,786 words, and *MPQA* (Wilson et al., 2005) has about 8,000 words.

There have been efforts to build sentiment polarity lexicons automatically. This includes midsize ones such as SentiWordNet (Baccianella et al., 2010; Esuli and Sebastiani, 2006), with 117K+ entries, annotated via projection of sentiment labels to WordNet synsets,[1] as well as large-scale ones such as the *Hashtag Sentiment Lexicon* and *Sentiment140* (Mohammad et al., 2013) with 680K and 1.2M+ entries, respectively.

SemEval-2015 Task 4 (Rosenthal et al., 2015) and SemEval-2016 Task 7 (Kiritchenko et al., 2016) featured a task on predicting the out-of-context sentiment intensity of words and phrases. Classic work on sentiment analysis lexicon construction and on phrase-level sentiment analysis includes Turney (2002); Wilson et al. (2005, 2009).

Ordinal Classification The above SemEval tasks were offered in different granularities, for example, 2-way (POSITIVE, NEGATIVE), 3-way (POSITIVE, NEUTRAL, NEGATIVE), 4-way (POSITIVE, NEUTRAL, NEGATIVE, OBJECTIVE), 5-way (HIGHLY POSITIVE, POSITIVE, NEUTRAL, NEGATIVE, HIGHLYNEGATIVE), and sometimes even 11-way (Ghosh et al., 2015). Note that most of them can be seen as ordinal scales, for example, −2, −1, 0, 1, and 2 for the 5-point scale. This changes the machine learning task as not all mistakes are equal anymore, for example, misclassifying a HIGHLYNEGATIVE example as HIGHLYPOSITIVE is a bigger mistake than misclassifying it as NEGATIVE. From a machine learning perspective, this is an *ordinal classification*, a reformulation that was explored in task 4 at SemEval-2016 and SemEval-2017 (Nakov et al., 2016b; Rosenthal et al., 2017).

Quantification Practical applications are not interested in the sentiment of a *specific tweet*. Rather, they look at estimating the *prevalence* of positive and negative tweets about a given topic in a set of tweets, for example, from some time interval, which is known as *quantification*. Note that quantification is not a mere by-product of classification, and a good quantifier is not necessarily a good classifier (Forman, 2008). Moreover, in case of multiple labels on an ordinal scale, we have yet another machine learning problem: *ordinal quantification*. Both versions of quantification were explored in task 4 at SemEval-2016 and SemEval-2017 (Nakov et al., 2016b; Rosenthal et al., 2017).

[1] In WordNet, words are grouped into sets of cognitive synonyms (synsets), each expressing a distinct concept. An example synset is {*good, right, ripe*}.

Specialized Sentiment Analysis One example is *sentiment analysis in the financial domain*, for example, predicting whether a given piece of text expresses bullish (optimistic; believing that the stock price will increase) or bearish (pessimistic; believing that the stock price will decline) sentiment associated with companies and stocks, as in SemEval-2017 Task 5 (Cortis et al., 2017). Another example is *sentiment analysis for code-mixed messages*, for example, of Hinglish and Spanglish, as in SemEval-2020 task 9 (Patwa et al., 2020). There has also been work on detecting whether an *explicit/implicit event is pleasant or unpleasant*, as in SemEval-2015 task 9 (Russo et al., 2015). Finally, there is *multi-modal sentiment analysis*, for example, sentiment analysis of memes, which can combine images with text, as in SemEval-2020 task 8 (Sharma et al., 2020).

3.2 Other Subjectivity Tasks

Emotion Detection This task asks to detect emotions such as *anger*, *anticipation*, *disgust*, *fear*, *joy*, *love*, *optimism*, *pessimism*, *sadness*, *surprise*, and *trust*, together with their intensity. Examples of joy, sadness, and anger are shown in (9)–(11). The inventory of emotions is typically inspired by the Plutchik's wheel of emotion (Plutchik, 1980). The task could target social media messages as in SemEval-2018 Task 1 (Mohammad et al., 2018) and in WASSA-2017 (Mohammad and Bravo-Marquez, 2017), news headlines as in SemEval-2007 Task 14 (Strapparava and Mihalcea, 2007), or dialog as in SemEval-2019 Task 3 (Chatterjee et al., 2019). A related recent survey covers emotion recognition in conversations (Poria et al., 2019). A related task is *emoji prediction*, for example, predicting which emojis would be suitable to use in a given tweet, as in SemEval-2018 Task 2 (Barbieri et al., 2018).

(9) JOY: *It's meant to be!! #happy #happy*
(10) SADNESS: *Feel so grim + ugly atm*
(11) ANGER: *My blood is boiling*

Stance Detection A related task is that of *stance detection*. The goal here is to determine whether the author of a piece of text is in favor of, against, or neutral with respect to a proposition or a target. For example, in (12), the author has a negative stance with respect to the proposition *women have the right to abortion*, even though the target is not mentioned at all. Similarly, in (13) there is a negative stance with respect to *Mitt Romney*, which is an implicit positive stance for *Barack Obama* (a tweet from the 2012 US presidential campaign).

(12) *A foetus has rights too! Make your voice heard.*
(Target: *women have the right to abortion* → AGAINST)

(13) *All Mitt Romney cares about is making money for the rich.*
(Target: *Barack Obama* → INFAVOR)

Stance detection was featured as task 6 at SemEval-2016 (Mohammad et al., 2016). The task was also an integral part of the Fake News Challenge (Hanselowski et al., 2018) as well as of the FEVER Shared Task on Fact Extraction and VERification (Thorne et al., 2018), and of the RumourEval tasks at SemEval in 2017 and 2019 (Derczynski et al., 2017; Gorrell et al., 2019). See also Küçük and Can (2020) for a recent survey on stance detection, and Mohtarami et al. (2018) and Nie et al. (2019) for some interesting recent work.

Abusive Language Detection This includes tasks such as detecting the use of offensive language and its target (e.g., a group or an individual), as in the OffensEval task at SemEval in 2019–2020 (Zampieri et al., 2019, 2020), which was offered in English, Arabic, Danish, Greek, and Turkish. It further includes *hate speech detection*, as in SemEval-2019 task 5 (Basile et al., 2019). Other related tasks include HASOC 2019 (Mandl et al., 2019) for English, German, and Hindi, HatEval 2019 (Basile et al., 2019b) for English and Spanish, GermEval 2019 for German (Struß et al., 2019), and TRAC 2020 (Kumar et al., 2018) for Bengali, Hindi, and English. Typically, small manually annotated datasets are used for training, but recently large-scale semi-supervised datasets have come into use (Rosenthal et al., 2020). See Vidgen and Derczynski (2020) for a recent survey on abusive language detection.

Political Bias Detection Also known as political ideology detection. Examples (14)–(16) show left-, center- and right-biased news titles, respectively. The analysis can be done at different granularity levels, for example, sentence-level (Baly et al., 2018, 2020b; Dinkov et al., 2019; Saez-Trumper et al., 2013), article-level (Baly et al., 2020a; Kulkarni et al., 2018), or source-level (Baly et al., 2018, 2019, 2020c). The bias can also be measured with respect to a specific polarizing topic (Darwish et al., 2020), for example, *climate change*, *gun control*, *immigration*, *Brexit*, *COVID-19*, and so on. Finally, bias can be characterized in terms of coverage, omission, geography (Saez-Trumper et al., 2013). See Baeza-Yates (2018) and Pitoura et al. (2018) for a discussion on bias on the Web.

(14) LEFT: *UN Says Climate Genocide is Coming. It's Actually Worse Than That*

(15) CENTER: *New study helps improve accuracy of future climate change predictions*
(16) RIGHT: *Doubling Down on Global Warming Alarmism*

Hyper-Partisanship Detection When bias gets extreme, for example, extreme left or extreme right, we can talk about *hyper-partisanship*, which is defined as blind, prejudiced, or unreasoning allegiance to one party, faction, cause, or person. The task of hyper-partisanship detection was featured at SemEval-2019 Task 4 (Kiesel et al., 2019). It has been argued that it is hard to tell apart computationally extreme-left from extreme-right bias, but that it is easier to distinguish hyper-partisan bias from typical left/center/right bias; see Potthast et al. (2018) for more details.

Propaganda Detection Propaganda aims at influencing people's mindset with the purpose of advancing a specific agenda. Influencing opinions is achieved through a number of rhetorical and psychological techniques, ranging from leveraging on emotions – such as using *loaded language*, *flag waving*, *appeal to authority*, *slogans*, and *clichés* – to using logical fallacies – such as *straw men*, *red herring*, *black-and-white fallacy*, *bandwagon*, and *whataboutism*. Examples are shown in (17)–(19).

(17) LOADED LANGUAGE: ***Outrage*** *as Donald Trump suggests injecting disinfectant to kill virus.*
(18) APPEAL TO FEAR: *Coronavirus could kill more Americans than some wars*
(19) BANDWAGON: *He tweeted,* ***"EU no longer considers #Hamas a terrorist group. Time for US to do same."***

Note that the definition of propaganda is orthogonal to that of *disinformation* (see in subsequent text). Note also that propaganda can hook to claims that are either true or false, and it can be either harmful or harmless (even good). In practice, propaganda and disinformation are used jointly to weaponize social media. Traditionally, research in propaganda detection was done at the article level (Barrón-Cedeño et al., 2019; Rashkin et al., 2017). More recent work has focused on detecting the use of specific propaganda techniques (Da San Martino et al., 2019b), which was also the topic of SemEval-2020 task 11 (Da San Martino et al., 2020a) as well as of a shared task at NLP4IF-2020 (Da San Martino et al., 2019a). For more detail, see the survey in Da San Martino et al. (2020b) and the special issue of the Big Data journal (Bolsover and Howard, 2017).

Disinformation and Factuality Detection Disinformation is defined as information that is both (*i*) *false* and (*ii*) *intents to harm*. The often-ignored aspect (*ii*) is the real reason why society started worrying back in 2016 about the rise of "fake news," namely because the news got *weaponized*. Yet, most research has focused on the veracity aspect as it is much easier to tackle, for example, there have been tasks on *rumor detection and verification* at SemEval in 2017 and 2019 (Derczynski et al., 2017; Gorrell et al., 2019), and on *fact-checking claims in community question answering forums* at SemEval-2019 (Mihaylova et al., 2019). Other related tasks include the CLEF 2018–2020 CheckThat! lab on idetification and verification of claims in political debates and in social media (Barrón-Cedeño et al., 2019; Elsayed et al., 2019; Nakov et al., 2018), and the FEVER-2018 task on Fact Extraction and VERification (Thorne et al., 2018). Interesting surveys and influential research include Lazer et al. (2018); Li et al. (2016); Shu et al. (2017); Thorne and Vlachos (2018); Vosoughi et al. (2018), and there was a special journal issue of ACM Trans. Inf. Syst. (Papadopoulos et al., 2016).

Argumentation Mining Argumentation mining is a very active research area with an annual specialized workshop. The task includes subtasks such as (*i*) identifying argumentation in text, (*ii*) characterizing the arguments as supporting or attacking, and (*iii*) understanding what licenses the argumentation. The latter subtask was featured as SemEval-2018 Task 12 (Habernal et al., 2018). A related task is that of *suggestion mining*, SemEval-2019 task 9 (Negi et al., 2019), which asks to extract subjective suggestions from reviews and forums, as shown in examples (20)–(21).

(20) SUGGESTION: *Be sure to specify a room at the back of the hotel.*
(21) NON-SUGGESTION: *It fails with a uninformative message indicating deployment failed.*

Important surveys and influential publications on argumentation mining include Lippi and Torroni (2016); Peldszus and Stede (2013); and Stede and Schneider (2018).

3.3 Machine Learning Models

Below, we discuss machine learning models and representations for sentiment analysis as a representative task for subjective language detection. The other tasks we discussed earlier use similar techniques.

Preprocessing Standard preprocessing steps for sentiment analysis include tokenization, stemming, lemmatization, stop-word removal, and part-of-speech tagging. As sentiment analysis is often done on noisy user-generated content, for example, from community forums or from social media, researchers often use Twitter-specific NLP tools such as specialized part-of-speech, named entity taggers, syntactic parsers, and so on. (Gimpel et al., 2011; Kong et al., 2014; Ritter et al., 2011). Moreover, when processing tweets, it is common to substitute/remove URLs, user mentions, hashtags, and emoticons, and also to perform some spelling correction, elongation normalization, abbreviation lookup, and punctuation removal.

Negation Handling Special handling is often done for negation. The most popular approach is to transform any word that appears in a negation context by adding a suffix *_NEG* to it, for example, *good* would become *good_NEG* (Das and Chen, 2007; Pang et al., 2002). A negated context is typically defined as a text span between a negation word, for example, *no, not, shouldn't*, and a punctuation mark or the end of the message. Alternatively, one could flip the polarity of sentiment words, for example, the positive word *good* would become negative when negated. It has also been argued (Zhu et al., 2014) that negation affects different words differently, and thus it was also proposed to build and use special sentiment polarity lexicons for words in negation contexts (Kiritchenko et al., 2014).

Features Traditionally, systems for Sentiment Analysis on Twitter have relied on handcrafted features derived from word-level (e.g., *great*, *freshly roasted coffee*, *becoming president*) and character-level *n*-grams (e.g., *bec*, *beco*, *comin*, *oming*), stems (e.g., *becom*), lemmata (e.g., *become*, *roast*), punctuation (e.g., exclamation and question marks), part-of-speech tags (e.g., adjectives, adverbs, verbs, nouns), word clusters (e.g., *probably*, *probly*, and *maybe* could be collapsed to the same word cluster), and Twitter-specific encodings such as emoticons (e.g., *;)*, *:D*), hashtags (*#Brexit*), user tags (e.g., *@allenai_org*), abbreviations (e.g., *RT*, *BTW*, *F2F*, *OMG*), elongated words (e.g., *soooo*, *yaayyy*), use of capitalization (e.g., proportion of ALL CAPS words), URLs, and so on. The most important features are those based on the presence of words and phrases in sentiment polarity lexicons with positive/negative scores; examples of such features include number of positive terms, number of negative terms, ratio of the number of positive terms to the number of positive+negative terms, ratio of the number of negative terms to the number of positive+negative terms, sum of all positive scores, sum of all negative scores, sum of all scores, and so on. More recently, with the emergence of word embeddings, they have been used as features as well, typically as part of a complex neural network architecture.

Supervised Learning Traditionally, the above features were fed into classifiers such as Maximum Entropy (MaxEnt) and Support Vector Machines

(SVM) with various kernels. However, observation over the SemEval Twitter sentiment task in recent years shows growing interest in, and by now clear dominance of methods based on deep learning. A number of neural network architectures have been applied to sentiment analysis, including convolutional neural networks (Kim, 2014), recurrent neural networks (Liu et al., 2015; Tang et al., 2016), and recursive neural networks (Socher et al., 2013; Tai et al., 2015). In particular, the best-performing systems at SemEval-2015 and SemEval-2016 used deep convolutional networks (Deriu et al., 2016; Severyn and Moschitti, 2015b). More recently, state-of-the-art performance was achieved using large-scale pretrained language models such as ELMo (Peters et al., 2018), BERT (Devlin et al., 2019), and RoBERTa (Liu et al., 2019).

Semi-Supervised Learning We should note two things about the use of deep neural networks. First, they can often do quite well without the need for explicit feature modeling, as they can learn the relevant features in their hidden layers starting from the raw text. Second, they have too many parameters, and thus they require a lot of training data, orders of magnitude more than it is realistic to have manually annotated. A popular way to solve this latter problem is to use self-training, a form of semi-supervised learning, where first a model is trained on the available training data only, then this model is applied to make predictions on a large unannotated set of tweets, and finally, it is trained for a few more iterations on its own predictions.

This works because parts of the network, for example, with convolution or with LSTMs (dos Santos and Gatti, 2014; Severyn and Moschitti, 2015b; Wang et al., 2015), needs to learn something like a language model, that is, which word is likely to follow which one. Training these parts needs no labels. A similar idea is used in pre-trained language models such as ELMo, BERT, and RoBERTa, but there the pre-training is done in a task-agnostic way. Recently, it has been shown that self-training can improve overusing such pretrained models, by using a data augmentation method that computes task-specific query embeddings from labeled data to retrieve sentences from billions of unlabeled sentences, which are then used to train a teacher-student framework (Du et al., 2020).

Unsupervised Learning Fully unsupervised learning is not popular for sentiment analysis. Yet, some features used in sentiment analysis have been learned in an unsupervised way, for example, Brown clusters to generalize over words (Owoputi et al., 2012). Similarly, word embeddings are typically trained on raw tweets that have no annotation for sentiment (even though there is also work on sentiment-specific word embeddings (Tang et al., 2014), which has used distant supervision). More recently, this was replaced by large-scale pretrained Transformers such as BERT.

Distant Supervision Another way to benefit from large unannotated datasets is to use *distant supervision* (Marchetti-Bowick and Chambers, 2012). For example, one can annotate tweets for sentiment polarity based on whether they contain a positive or a negative emoticon. This results in noisy labels, which can be used to train a system (Severyn and Moschitti, 2015b) to induce sentiment-specific word embeddings (Tang et al., 2014) and sentiment-polarity lexicons (Mohammad et al., 2013).

Bootstrapping Sentiment Polarity Lexicons Traditionally, sentiment polarity lexicons have been the most commonly used resource for sentiment analysis. Thus, various approaches have been proposed in the literature to build such lexicons automatically. The dominant approach is that of Turney (2002). It starts with a small set of seed positive (e.g., *excellent*) and negative words (e.g., *bad*), and then uses these words to induce sentiment polarity orientation for new words in a large unannotated set of texts (in his case, product reviews). The idea is that words that co-occur in the same text with positive seed words are likely to be positive, while such that tend to co-occur with negative words are likely to be negative. To quantify this intuition, Turney defined the notion of sentiment orientation (SO) for a term w as follows:

$$SO(w) = pmi(w, pos) - pmi(w, neg)$$

In this formula, PMI is the pointwise mutual information, pos and neg are placeholders standing for any of the seed positive and negative terms, respectively, and w is a target word/phrase from the large unannotated set of texts.

A positive/negative value for $SO(w)$ indicates positive/negative polarity for the word w, and its magnitude shows the corresponding sentiment strength. In turn, $pmi(w, pos) = \frac{P(w,pos)}{P(w)P(pos)}$, where $P(w, pos)$ is the probability to see w with any of the seed positive words, $P(w)$ is the probability to see w, and $P(pos)$ is the probability to see any of the seed positive words; $pmi(w, neg)$ is defined similarly.

Pointwise mutual information is a notion from information theory: the *mutual information* of two random variables A and B is the amount of information (in units such as bits) obtained about A through B (Church and Hanks, 1990). Let a and b be two values from the sample space of A and B, respectively. The *pointwise mutual information* between a and b is defined as follows:

$$pmi(a;b) = \log \frac{P(A = a, B = b)}{P(A = a) \cdot P(B = b)} = \log \frac{P(A = a | B = b)}{P(A = a)} \tag{3.1}$$

In his experiments, Turney used five positive and five negative words as seeds (Turney, 2002). His PMI-based approach further served as the basis for the creation of two large-scale automatic lexicons for sentiment analysis in Twitter for English, initially developed by NRC for their participation in SemEval-2013 (Mohammad et al., 2013). The *Hashtag Sentiment Lexicon* uses as seeds hashtags containing 32 positive and 36 negative words, for example, `#happy` and `#sad`. Similarly, the *Sentiment140* lexicon uses smileys as seed indicators for positive and negative sentiment, for example, `:)`, `:-)`, and `:))` as positive seeds, and `:(` and `:-(` as negative ones.

An alternative approach to lexicon induction has been proposed (Severyn and Moschitti, 2015a), which, instead of using PMI, assigns positive/negative labels to the unlabeled tweets (based on the seeds), and then trains an SVM classifier on them, using word n-grams as features. These n-grams are then used as lexicon entries (words and phrases) with the learned classifier weights as polarity scores. Finally, it has been shown that sizable further performance gains can be obtained by starting with midsized seeds, that is, hundreds of words and phrases (Jovanoski et al., 2016).

Nowadays, the importance of such lexicons is diminished due to the emergence of large-scale pre-trained Transformers such as BERT.

3.4 Conclusion and Recommended Readings

We have explored a number of tasks tackling subjective language, with focus on sentiment analysis. We expect the quest for new interesting tasks to continue, with competitions such as those at SemEval as the engine of innovation, as they not only perform head-to-head comparisons but also create databases and tools that enable follow-up research for many years afterward.

The interested reader can learn more about how subjective language is treated in computational linguistics by checking some of the abovementioned references. A great starting point are surveys, for example, on sentiment analysis (Liu and Zhang, 2012; Pang and Lee, 2008), on emotion recognition (Poria et al., 2019), on stance detection (Küçük and Can, 2020), on abusive language detection (Vidgen and Derczynski, 2020), on the biases on the Web (Baeza-Yates, 2018), on propaganda detection (Da San Martino et al., 2020b), on disinformation and fact-checking (Lazer et al., 2018; Li et al., 2016; Shu et al., 2017; Thorne and Vlachos, 2018; Vosoughi et al., 2018), and on argumentation mining (Lippi and Torroni, 2016; Peldszus and Stede, 2013; Stede and Schneider, 2018). We further recommend special issues of journals on the topic (Balahur et al., 2014; Bolsover and Howard, 2017; Papadopoulos et al., 2016), as well as the overview papers for the above-mentioned shared tasks.

References

Baccianella, S., Esuli, A., and Sebastiani, F. 2010. SentiWordNet 3.0: An enhanced lexical resource for sentiment analysis and opinion mining. *Proceedings of the Seventh International Conference on Language Resources and Evaluation*. LREC '10. Valletta: European Language Resources Association.

Baeza-Yates, R. 2018. Bias on the Web. *Communications of the ACM*, **61**(6), 54–61.

Balahur, A., Mihalcea, R., and Montoyo, A. 2014. Computational approaches to subjectivity and sentiment analysis: Present and envisaged methods and applications. *Computer Speech & Language*, **28**(1), 1–6.

Baly, R., Karadzhov, G., Alexandrov, D., Glass, J., and Nakov, P. 2018. Predicting factuality of reporting and bias of news media sources. Pages 3528–3539 of: *Proceedings of the 2018 Conference on Empirical Methods in Natural Language Processing*. EMNLP '18. Brussels: Association for Computational Linguistics.

Baly, R., Karadzhov, G., Saleh, A., Glass, J., and Nakov, P. 2019. Multi-task ordinal regression for jointly predicting the trustworthiness and the leading political ideology of news media. Pages 2109–2116 of: *Proceedings of the 17th Annual Conference of the North American Chapter of the Association for Computational Linguistics: Human Language Technologies*. NAACL-HLT'2019. Brussels:Association for Computational Linguistics.

Baly, R., Martino, G. D. S., Glass, J., and Nakov, P. 2020a. We can detect your bias: Predicting the political ideology of news articles. *Proceedings of the 2020 Conference on Empirical Methods in Natural Language Processing*. EMNLP '20. Stroudsburg: Association for Computational Linguistics.

Baly, R., Karadzhov, G., An, J., Kwak, H., Dinkov, Y., Ali, A., Glass, J., and Nakov, P. 2020c. What was written vs. who read it: News media profiling using text analysis and social media context. *Proceedings of the 58th Annual Meeting of the Association for Computational Linguistics*. Stroudsburg: ACL '20.

Barbieri, F., Camacho-Collados, J., Ronzano, F., Espinosa-Anke, L., Ballesteros, M., Basile, V., Patti, V., and Saggion, H. 2018. SemEval 2018 Task 2: Multilingual emoji prediction. Pages 24–33 of: *Proceedings of the 12th International Workshop on Semantic Evaluation*. SemEval '18. New Orleans LA: Association for Computational Linguistics.

Barrón-Cedeño, A., Elsayed, T., Nakov, P., Da San Martino, G., Hasanain, M., Suwaileh, R., and Haouari, F. 2019. CheckThat! at CLEF 2020: Enabling the automatic identification and verification of claims on social media. Pages 499–507 of: *Proceedings of the 42d European Conference on Information Retrieval*. ECIR '20. Berlin: Springer.

Barrón-Cedeño, A., Da San Martino, G., Jaradat, I., and Nakov, P. 2019. Proppy: A system to unmask propaganda in online news. Pages 9847–9848 of: *Proceedings of the 33rd AAAI Conference on Artificial Intelligence*. AAAI '19. California: AAAI Press.

Basile, V., Bosco, C., Fersini, E., Nozza, D., Patti, V., Rangel Pardo, F. M., Rosso, P., and Sanguinetti, M. 2019a. SemEval-2019 Task 5: Multilingual detection of hate speech against immigrants and women in Twitter. Pages 54–63 of: *Proceedings of the 13th International Workshop on Semantic Evaluation*. SemEval '19. Stroudsburg: Association for Computational Linguistics.

Bollen, J., Mao, H., and Zeng, X.-J. 2011. Twitter mood predicts the stock market. *Journal of Computational Science*, **2**(1), 1–8.

Bolsover, G., and Howard, P. 2017. Computational propaganda and political big data: Moving toward a more critical research agenda. *Big Data*, **5**(4), 273–276.

Borge-Holthoefer, J., Magdy, W., Darwish, K., and Weber, I. 2015. Content and network dynamics behind Egyptian political polarization on Twitter. Pages 700–711 of: *Proceedings of the 18th ACM Conference on Computer Supported Cooperative Work and Social Computing*. CSCW '15. Vancouver: ACM.

Burton, S. and Soboleva, A. 2011. Interactive or reactive? Marketing with Twitter. *Journal of Consumer Marketing*, **28**(7), 491–499.

Chatterjee, A., Narahari, K. N., Joshi, M., and Agrawal, P. 2019. SemEval-2019 Task 3: EmoContext contextual emotion detection in text. Pages 39–48 of: *Proceedings of the 13th International Workshop on Semantic Evaluation*. SemEval '19. Stroudsburg: Association for Computational Linguistics.

Church, K. W., and Hanks, P. 1990. Word association norms, mutual information, and lexicography. *Computational Linguistics Journal*, **16**(1), 22–29.

Cortis, K., Freitas, A., Daudert, T., Huerlimann, M., Zarrouk, M., Handschuh, S., and Davis, B. 2017 (August). SemEval-2017 Task 5: Fine-grained sentiment analysis on financial microblogs and news. Pages 519–535 of: *Proceedings of the 11th International Workshop on Semantic Evaluation*. SemEval '17. Stroudsburg: Association for Computational Linguistics.

Da San Martino, G., Barron-Cedeno, A., and Nakov, P. 2019a. Findings of the NLP4IF-2019 shared task on fine-grained propaganda detection. Pages 162–170 of: *Proceedings of the 2nd Workshop on NLP for Internet Freedom (NLP4IF): Censorship, Disinformation, and Propaganda*. NLP4IFEMNLP '19. Stroudsburg: Association for Computational Linguistics.

Da San Martino, G., Yu, S., Barrón-Cedeño, A., Petrov, R., and Nakov, P. 2019b. Fine-grained analysis of propaganda in news articles. Pages 5636–5646 of: *Proceedings of the 2019 Conference on Empirical Methods in Natural Language Processing and the 9th International Joint Conference on Natural Language Processing*. Vancouver: Association for Computational Linguistics.

Da San Martino, G., Barrón-Cedeño, A., Wachsmuth, H., Petrov, R., and Nakov, P. 2020a. SemEval-2020 Task 11: Detection of propaganda techniques in news articles. *Proceedings of the International Workshop on Semantic Evaluation*. SemEval '20. Vancouver: Association for Computational Linguistics.

Da San Martino, G., Cresci, S., Barrón-Cedeño, A., Yu, S., Di Pietro, R., and Nakov, P. 2020b. A survey on computational propaganda detection. *Proceedings of the 29th International Joint Conference on Artificial Intelligence and the 17th Pacific Rim International Conference on Artificial Intelligence*. IJCAI-PRICAI '20.

Darwish, K., Aupetit, M., Stefanov, P., and Nakov, P. 2020. Unsupervised user stance detection on Twitter. Pages 141–152 of: *Proceedings of the International AAAI Conference on Web and Social Media*. ICWSM '20. Atlanta: AAAI Press.

Das, S. R. and Chen, M. Y. 2007. Yahoo! for Amazon: Sentiment extraction from small talk on the Web. *Management Science Journal*, **53**(9), 1375–1388.

Derczynski, L., Bontcheva, K., Liakata, M., Procter, R., Wong Sak Hoi, G., and Zubiaga, A. 2017 (August). SemEval-2017 Task 8: RumourEval: Determining rumour veracity and support for rumours. Pages 69–76 of: *Proceedings of the 11th International Workshop on Semantic Evaluation*. SemEval '17. Vancouver: Association for Computational Linguistics.

Deriu, J., Gonzenbach, M., Uzdilli, F., Lucchi, A., De Luca, V., and Jaggi, M. 2016. SwissCheese at SemEval-2016 Task 4: Sentiment classification using an ensemble of convolutional neural networks with distant supervision. Pages 1124–1128 of: *Proceedings of the 10th International Workshop on Semantic Evaluation*. SemEval '16. San Diego: Association for Computational Linguistics.

Devlin, J., Chang, M.-W., Lee, K., and Toutanova, K. 2019. BERT: Pre-training of deep bidirectional transformers for language understanding. Pages 4171–4186 of: *Proceedings of the 2019 Conference of the North American Chapter of the Association for Computational Linguistics: Human Language Technologies*. NAACL-HLT '19. Brussels: Association for Computational Linguistics.

Dinkov, Y., Ali, A., Koychev, I., and Nakov, P. 2019. Predicting the leading political ideology of YouTube channels using acoustic, textual, and metadata information. Pages 501–505 of: Kubin, G., and Kacic, Z. (eds), *Proceedings of the 20th Annual Conference of the International Speech Communication Association*. Graz: International Speech Communication Association.

Dodds, P. S., Harris, K. D., Kloumann, I. M., Bliss, C. A., and Danforth, C. M. 2011. Temporal patterns of happiness and information in a global social network: Hedonometrics and Twitter. *PLoS ONE*, **6**(12).

dos Santos, C., and Gatti, M. 2014. Deep convolutional neural networks for sentiment analysis of short texts. Pages 69–78 of: *Proceedings of the 25th International Conference on Computational Linguistics*. COLING '14. Dublin: Dublin City University and Association for Computational Linguistics.

Du, J., Grave, E., Gunel, B., Chaudhary, V., Celebi, O., Auli, M., Stoyanov, V., and Conneau, A. 2020. Self-training improves pre-training for natural language understanding. *Proceedings of the Annual Conference of the North American Chapter of the Association for Computational Linguistics: Human Language Technologies*. NAACL-HLT'2021. Stroudsburg: Association for Computational Linguistics.

Elsayed, T., Nakov, P., Barrón-Cedeño, A., Hasanain, M., Suwaileh, R., Da San Martino, G., and Atanasova, P. 2019. Overview of the CLEF-2019 CheckThat!: Automatic identification and verification of claims. Pages of 301–321: Crestani, F., Braschler, M., Savoy, J., Rauber, A., Muller, H., Losada, D. E., Burki, G. H., Cappellato, L., and Ferro, N. *Experimental IR meets multilinguality, multimodality, and interaction*. LNCS. Lugano: Springer.

Esuli, A., and Sebastiani, F. 2006. SENTIWORDNET: A publicly available lexical resource for opinion mining. Pages 417–422 of: *Proceedings of the International Conference on Language Resources and Evaluation*. LREC '06. Genoa: European Language Resources Association.

Forman, G. 2008. Quantifying counts and costs via classification. *Data Mining and Knowledge Discovery*, **17**(2), 164–206.

Ghosh, A., Li, G., Veale, T., Rosso, P., Shutova, E., Barnden, J., and Reyes, A. 2015. SemEval-2015 Task 11: Sentiment analysis of figurative language in Twitter. Pages 470–478 of: *Proceedings of the 9th International Workshop on Semantic Evaluation*. SemEval '15. Denver: for Computational Linguistics.

Gimpel, K., Schneider, N., O'Connor, B., Das, D., Mills, D., Eisenstein, J., Heilman, M., Yogatama, D., Flanigan, J., and Smith, N. A. 2011. Part-of-speech tagging for Twitter: Annotation, features, and experiments. Pages 42–47 of: *Proceedings of the 49th Annual Meeting of the Association for Computational Linguistics: Human Language Technologies*. ACL-HLT '11. Portland: Association for Computational Linguistics.

Gorrell, G., Kochkina, E., Liakata, M., Aker, A., Zubiaga, A., Bontcheva, K., and Derczynski, L. 2019. SemEval-2019 Task 7: RumourEval, determining rumour veracity and support for rumours. Pages 845–854 of: *Proceedings of the 13th International Workshop on Semantic Evaluation*. SemEval '19. Minneapolis: Association for Computational Linguistics.

Habernal, I., Wachsmuth, H., Gurevych, I., and Stein, B. 2018. SemEval-2018 Task 12: The argument reasoning comprehension task. Pages 763–772 of: *Proceedings of the 12th International Workshop on Semantic Evaluation*. SemEval '18. New Orleans: Association for Computational Linguistics.

Hanselowski, A., P. V. S., A., Schiller, B., Caspelherr, F., Chaudhuri, D., Meyer, C. M., and Gurevych, I. 2018. A retrospective analysis of the fake news challenge stance-detection task. Pages 1859–1874 of: *Proceedings of the International Conference on Computational Linguistics*. COLING'2018. Santa Fe: Association for Computational Linguistics

Hu, M., and Liu, B. 2004. Mining and summarizing customer reviews. Pages 168–177 of: *Proceedings of the 10th ACM SIGKDD International Conference on Knowledge Discovery and Data Mining*. KDD '04. Seattle: Association for Computing Machinery.

Java, A., Song, X., Finin, T., and Tseng, B. 2007. Why we Twitter: Understanding microblogging usage and communities. Pages 56–65 of: *Proceedings of the 9th WebKDD and 1st SNA-KDD 2007 Workshop on Web Mining and Social Network Analysis*. WebKDD/SNA-KDD'2007. San Jose: Association for Computing Machinery.

Jovanoski, D., Pachovski, V., and Nakov, P. 2016 (December). On the impact of seed words on sentiment polarity lexicon induction. *Proceedings of the 26th International Conference on Computational Linguistics*. COLING '16. Osaka: The COLING 2016 Organizing Committee.

Kaya, M., Fidan, G., and Toroslu, I. H. 2013. Transfer learning using Twitter data for improving sentiment classification of Turkish political news. Pages 139–148 of: *Proceedings of the 28th International Symposium on Computer and Information Sciences*. ISCIS '13. Berlin: Springer.

Kiesel, J., Mestre, M., Shukla, R., Vincent, E., Adineh, P., Corney, D., Stein, B., and Potthast, M. 2019. SemEval-2019 Task 4: Hyperpartisan news detection. Pages 829–839 of: *Proceedings of the 13th International Workshop on Semantic Evaluation*. Minneapolis MN: Association for Computational Linguistics.

Kim, Y. 2014. Convolutional neural networks for sentence classification. Pages 1746–1751 of: *Proceedings of the 2014 Conference on Empirical Methods in Natural Language Processing*. EMNLP '14. Doha: Association for Computational Linguistics.

Kiritchenko, S., Zhu, X., and Mohammad, S. M. 2014. Sentiment analysis of short informal texts. *Journal of Artificial Intelligence Research*, **50**, 723–762.

Kiritchenko, S., Mohammad, S. M., and Salameh, M. 2016. SemEval-2016 Task 7: Determining sentiment intensity of English and Arabic phrases. *Proceedings of the 10th International Workshop on Semantic Evaluation*. SemEval '16. San Diego: Association for Computational Linguistics.

Kong, L., Schneider, N., Swayamdipta, S., Bhatia, A., Dyer, C., and Smith, N. A. 2014. A dependency parser for tweets. Pages 1001–1012 of: *Proceedings of the 2014 Conference on Empirical Methods in Natural Language Processing*. EMNLP '14. Doha: Association for Computational Linguistics.

Kouloumpis, E., Wilson, T., and Moore, J. 2011. Twitter sentiment analysis: The good the bad and the OMG! Pages 538–541 of: *Proceedings of the Fifth International Conference on Weblogs and Social Media*. ICWSM '11. Barcelona: AAAI Press.

Küçük, D., and Can, F. 2020. Stance detection: A survey. *ACM Computing Surveys Journal*, **53**(1), 1–37.

Kulkarni, V., Ye, J., Skiena, S., and Wang, W. Y. 2018. Multi-view models for political ideology detection of news articles. Pages 3518–3527 of: *Proceedings of the Conference on Empirical Methods in Natural Language Processing*. EMNLP '18. Brussels: Association for Computational Linguistics.

Kumar, R., Ojha, A. K., Malmasi, S., and Zampieri, M. 2018. Evaluating aggression identification in social media. *Proceedings of the Second Workshop on Trolling, Aggression and Cyberbulling*. TRAC'2020. Marseille: European Language Resources Association.

Kwak, H., Lee, C., Park, H., and Moon, S. 2010. What is Twitter, a social network or a news media? Pages 591–600 of: *Proceedings of the 19th International Conference on World Wide Web*. WWW '10. Raleigh: Association for Computing Machinery.

Lazer, D. M. J., Baum, M. A., Benkler, Y., Berinsky, A. J., Greenhill, K. M., Menczer, F., Metzge , M. J., Nyhan, B., Pennycook, G., Rothschild, D., Schudson, M., Sloman, S. A., Sunstein, C. R., Thorson, E. A., Watts, D. J., and Zittrain, J. L. 2018. The science of fake news. *Science*, **359**(6380), 1094–1096.

Li, Y., Gao, J., Meng, C., Li, Q., Su, L., Zhao, B., Fan, W., and Han, J. 2016. A survey on truth discovery. *ACM SIGKDD Explorations Newsletter*, **17**(2), 1–16.

Lippi, M., and Torroni, P. 2016. Argumentation mining: State of the art and emerging trends. *ACM Transactions on Internet Technology*, **16**(2), 1–25.

Liu, B. and Zhang, L. 2012. A survey of opinion mining and sentiment analysis. Pages 415–463 of: Aggarwal, C. C., and Zhai, C. X. (eds), *Mining text data*. Berlin: Springer.

Liu, P., Qiu, X., Chen, X., Wu, S., and Huang, X. 2015. Multi-timescale long short-term memory neural network for modelling sentences and documents. Pages 2326–2335 of: *Proceedings of the 2015 Conference on Empirical Methods in Natural Language Processing*. EMNLP '15. Lisbon: Association for Computational Linguistics.

Liu, Y., Ott, M., Goyal, N., Du, J., Joshi, M., Chen, D., Levy, O., Lewis, M., Zettlemoyer, L., and Stoyanov, V. 2019. *RoBERTa: A robustly optimized BERT pretraining approach*. arXiv preprint arXiv:1907.11692.

Mandl, T., Modha, S., Majumder, P., Patel, D., Dave, M., Mandlia, C., and Patel, A. 2019. Overview of the HASOC track at FIRE 2019: Hate speech and offensive content identification in Indo-European languages. *Proceedings of the 11th*

Forum for Information Retrieval Evaluation. FIRE'2019. Kolkata: Association for Computing Machinery.

Marchetti-Bowick, M., and Chambers, N. 2012. Learning for microblogs with distant supervision: Political forecasting with Twitter. Pages 603–612 of: *Proceedings of the 13th Conference of the European Chapter of the Association for Computational Linguistics*. EACL '12. Avignon: Association for Computational Linguistics.

Mihaylova, T., Karadzhov, G., Atanasova, P., Baly, R., Mohtarami, M., and Nakov, P. 2019. SemEval-2019 Task 8: Fact checking in community question answering forums. Pages 860–869 of: *Proceedings of the 13th International Workshop on Semantic Evaluation*. SemEval '19. Minneapolis: Association for Computational Linguistics.

Mohammad, S. and Bravo-Marquez, F. 2017. WASSA-2017 shared task on emotion intensity. Pages 34–49 of: *Proceedings of the 8th Workshop on Computational Approaches to Subjectivity, Sentiment and Social Media Analysis*. WASSA '17. Copenhagen: Association for Computational Linguistics.

Mohammad, S. Kiritchenko, S., and Zhu, X. 2013. NRC-Canada: Building the state-of-the-art in sentiment analysis of tweets. Pages 321–327 of: *Proceedings of the Second Joint Conference on Lexical and Computational Semantics (*SEM), Volume 2: Proceedings of the Seventh International Workshop on Semantic Evaluation*. SemEval '13. Atlanta: Association for Computational Linguistics.

Mohammad, S., Bravo-Marquez, F., Salameh, M., and Kiritchenko, S. 2018. SemEval-2018 Task 1: Affect in tweets. Pages 1–17 of: *Proceedings of the 12th International Workshop on Semantic Evaluation* SemEval '18. New Orleans: Association for Computational Linguistics.

Mohammad, S. M., Kiritchenko, S., Sobhani, P., Zhu, X., and Cherry, C. 2016. SemEval-2016 Task 6: Detecting stance in tweets. Pages 31–41 of: *Proceedings of the 10th International Workshop on Semantic Evaluation*. SemEval '16. San Diego: Association for Computational Linguistics.

Mohtarami, M., Baly, R., Glass, J., Nakov, P., Màrquez, L., and Moschitti, A. 2018. Automatic stance detection using end-to-end memory networks. Pages 767–776 of: *Proceedings of the 16th Annual Conference of the North American Chapter of the Association for Computational Linguistics: Human Language Technologies*. NAACL-HLT '18. New Orleans: Association for Computational Linguistics.

Nakov, P., Rosenthal, S., Kozareva, Z., Stoyanov, V., Ritter, A., and Wilson, T. 2013. SemEval-2013 Task 2: Sentiment analysis in Twitter. Pages 312–320 of: *Proceedings of the Second Joint Conference on Lexical and Computational Semantics (*SEM), Volume 2: Proceedings of the Seventh International Workshop on Semantic Evaluation*. SemEval '13. Atlanta: Association for Computational Linguistics.

Nakov, P., Rosenthal, S., Kiritchenko, S., Mohammad, S. M., Kozareva, Z., Ritter, A., Stoyanov, V., and Zhu, X. 2016a. Developing a successful SemEval task in sentiment analysis of Twitter and other social media texts. *Language Resources and Evaluation*, **50**(1), 35–65.

Nakov, P., Màrquez, L., Moschitti, A., Magdy, W., Mubarak, H., Freihat, A. A., Glass, J., and Randeree, B. 2016b. SemEval-2016 Task 3: Community question answering.

Pages 525–545 of: *Proceedings of the 10th International Workshop on Semantic Evaluation*. SemEval '16. San Diego: Association for Computational Linguistics.

Nakov, P., Barrón-Cedeño, A., Elsayed, T., Suwaileh, R., Màrquez, L., Zaghouani, W., Atanasova, P., Kyuchukov, S., and Da San Martino, G. 2018. Overview of the CLEF-2018 CheckThat! Lab on automatic identification and verification of political claims. Pages 372–387 of: *Proceedings of the Ninth International Conference of the CLEF Association: Experimental IR Meets Multilinguality, Multimodality, and Interaction*. Lecture Notes in Computer Science. Avignon: Springer.

Negi, S., Daudert, T., and Buitelaar, P. 2019. SemEval-2019 Task 9: Suggestion mining from online reviews and forums. Pages 877–887 of: *Proceedings of the 13th International Workshop on Semantic Evaluation*. SemEval '19. Minneapolis: Association for Computational Linguistics.

Nie, Y., Chen, H., and Bansal, M. 2019. Combining fact extraction and verification with neural semantic matching networks. Pages 6859–6866 of: *Proceedings of the 33rd AAAI Conference on Artificial Intelligence*. AAAI '19. Honolulu: AAAI Press.

O'Connor, B., Balasubramanyan, R., Routledge, B., and Smith, N. 2010. From tweets to polls: Linking text sentiment to public opinion time series. Pages 122–129 of: *Proceedings of the Fourth International Conference on Weblogs and Social Media*. ICWSM '10. Washington: AAAI Press.

Owoputi, O., Dyer, C., Gimpel, K., and Schneider, N. 2012. Part-of-speech tagging for Twitter: Word clusters and other advances. Technical report. CMU-ML-12-107. School of Computer Science, Carnegie Mellon University, Pittsburgh, PA.

Pang, B. and Lee, L. 2004. A sentimental education: Sentiment analysis using subjectivity summarization based on minimum cuts. Pages 271–278 of: *Proceedings of the 42nd Annual Meeting of the Association for Computational Linguistics*. ACL '04. Barcelona: Association for Computational Linguistics.

Pang, B., and Lee, L. 2008. Opinion mining and sentiment analysis. *Foundations and Trends in Information Retrieval*, **2**(1–2), 1–135.

Pang, B., Lee, L., and Vaithyanathan, S. 2002. Thumbs up?: Sentiment classification using machine learning techniques. Pages 79–86 of: *Proceedings of the Conference on Empirical Methods in Natural Language Processing*. EMNLP '02. Philadelphia: Association for Computational Linguistics.

Papadopoulos, S., Bontcheva, K., Jaho, E., Lupu, M., and Castillo, C. 2016. Overview of the special issue on trust and veracity of information in social media. *ACM Transactions on Information Systems*, **34**(3), 14:1–14:5.

Patwa, P., Aguilar, G., Kar, S., Pandey, S., PYKL, S., Garrette, D., Gambäck, B., Chakraborty, T., Solorio, T., and Das, A. 2020. SemEval-2020 Sentimix Task 9: Overview of SENTIment analysis of code-MIXed tweets. *Proceedings of the 14th International Workshop on Semantic Evaluation*. SemEval '20. Stroudsburg: Association for Computational Linguistics.

Peldszus, A., and Stede, M. 2013. From argument diagrams to argumentation mining in texts: A survey. *International Journal of Cognitive Informatics and Natural Intelligence*, **7**(1), 1–31.

Pennebaker, J. W., Francis, M. E., and Booth, R. J. 2001. *Linguistic inquiry and word count*. Mahwah, NJ: Lawerence Erlbaum Associates.

Peters, M., Neumann, M., Iyyer, M., Gardner, M., Clark, C., Lee, K., and Zettlemoyer, L. 2018. Deep contextualized word representations. Pages 2227–2237 of: *Proceedings of the 2018 Conference of the North American Chapter of the Association for Computational Linguistics: Human Language Technologies*. NAACL-HLT '18. New Orleans: Association for Computational Linguistics.

Pitoura, E., Tsaparas, P., Flouris, G., Fundulaki, I., Papadakos, P., Abiteboul, S., and Weikum, G. 2018. On measuring bias in online information. *ACM SIGMOD Record*, **46**(4), 16–21.

Plutchik, R. 1980. A general psychoevolutionary theory of emotion. *Theories of Emotion*, **1**, 3–31.

Pontiki, M., Papageorgiou, H., Galanis, D., Androutsopoulos, I., Pavlopoulos, J., and Manandhar, S. 2014. SemEval-2014 Task 4: Aspect based sentiment analysis. Pages 27–35 of: *Proceedings of the 8th International Workshop on Semantic Evaluation*. SemEval '14: Dublin: Association for Computational Linguistics.

Pontiki, M., Galanis, D., Papageorgiou, H., Manandhar, S., and Androutsopoulos, I. 2015. SemEval-2015 Task 12: Aspect based sentiment analysis. Pages 486–495 of: *Proceedings of the 9th International Workshop on Semantic Evaluation*. SemEval '15. Denver: Association for Computational Linguistics.

Pontiki, M., Galanis, D., Papageorgiou, H., Androutsopoulos, I., Manandhar, S., AL-Smadi, M., Al-Ayyoub, M., Zhao, Y., Qin, B., De Clercq, O., Hoste, V., Apidianaki, M., Tannier, X., Loukachevitch, N., Kotelnikov, E., Bel, N., Jiménez-Zafra, S. M., and Eryiğit, G. 2016. SemEval-2016 Task 5: Aspect based sentiment analysis. Pages 19–30 of: *Proceedings of the 10th International Workshop on Semantic Evaluation*. SemEval '16. San Diego: Association for Computational Linguistics.

Poria, S., Majumder, N., Mihalcea, R., and Hovy, E. H. 2019. Emotion recognition in conversation: Research challenges, datasets, and recent advances. *IEEE Access*, **7**, 100943–100953.

Potthast, M., Kiesel, J., Reinartz, K., Bevendorff, J., and Stein, B. 2018. A stylometric inquiry into hyperpartisan and fake news. Pages 231–240 of: *Proceedings of the 56th Annual Meeting of the Association for Computational Linguistics*. ACL '18. Melbourne: Association for Computational Linguistics.

Qureshi, M. A., O'Riordan, C., and Pasi, G. 2013. Clustering with error estimation for monitoring reputation of companies on Twitter. Pages 170–180 of: *Proceedings of the 9th Asia Information Retrieval Societies Conference*. AIRS '13. Berlin: Springer.

Rashkin, H., Choi, E., Jang, J. Y., Volkova, S., and Choi, Y. 2017. Truth of varying shades: Analyzing language in fake news and political fact-checking. Pages 2931–2937 of: *Proceedings of the Conference on Empirical Methods in Natural Language Processing*. Copenhagen: Association for Computational Linguistics.

Ritter, A., Clark, S., Mausam, and Etzioni, O. 2011. Named entity recognition in tweets: An experimental study. Pages 1524–1534 of: *Proceedings of the Conference on Empirical Methods in Natural Language Processing*. EMNLP '11. Edinburgh: Association for Computational Linguistics.

Rosenthal, S., Ritter, A., Nakov, P., and Stoyanov, V. 2014. SemEval-2014 Task 9: Sentiment analysis in Twitter. Pages 73–80 of: *Proceedings of the 8th International Workshop on Semantic Evaluation*. SemEval '14. Dublin: Association for Computational Linguistics.

Rosenthal, S., Nakov, P., Kiritchenko, S., Mohammad, S., Ritter, A., and Stoyanov, V. 2015. SemEval-2015 Task 10: Sentiment analysis in Twitter. Pages 450–462 of: *Proceedings of the 9th International Workshop on Semantic Evaluation.* SemEval '15. Vancouver: Association for Computational Linguistics.

Rosenthal, S., Farra, N., and Nakov, P. 2017. SemEval-2017 Task 4: Sentiment analysis in Twitter. Pages 502–518 of: *Proceedings of the 11th International Workshop on Semantic Evaluation.* SemEval '17. Vancouver: Association for Computational Linguistics.

Rosenthal, S., Atanasova, P., Karadzhov, G., Zampieri, M., and Nakov, P. 2020. A large-scale semi-supervised dataset for offensive language identification. Pages of 915–928: *Findings of the Association for Computational Linguistics.* ACL-IJCNLP 2021. Association for Computational Linguistics.

Russo, I., Caselli, T., and Strapparava, C. 2015. SemEval-2015 Task 9: CLIPEval implicit polarity of events. Pages 442–449 of: *Proceedings of the 9th International Workshop on Semantic Evaluation.* SemEval '15. Denver: Association for Computational Linguistics.

Saez-Trumper, D., Castillo, C., and Lalmas, M. 2013. Social media news communities: Gatekeeping, coverage, and statement bias. Pages 1679–1684 of: *Proceedings of the 22nd ACM International Conference on Information & Knowledge Management.* CIKM '13. San Francisco: Association for Computing Machinery.

Severyn, A., and Moschitti, A. 2015a. On the automatic learning of sentiment lexicons. Pages 1397–1402 of: *Proceedings of the 2015 Conference of the North American Chapter of the Association for Computational Linguistics: Human Language Technologies.* NAACL-HLT '15. Denver: Association for Computational Linguistics.

Severyn, A., and Moschitti, A. 2015b. Twitter sentiment analysis with deep convolutional neural networks. Pages 959–962 of: *Proceedings of the 38th International ACM SIGIR Conference on Research and Development in Information Retrieval.* SIGIR '15. Santiago: Association for Computing Machinery.

Sharma, C., Bhageria, D., Paka, W., Scott, W., PYKL, S., Das, A., Chakraborty, T., and Gambäck, B. 2020. SemEval-2020 Task 8: Memotion analysis-the visuo-lingual metaphor! *Proceedings of the 14th International Workshop on Semantic Evaluation.* SemEval '20. Stroudsburg: Association for Computational Linguistics.

Shu, K., Sliva, A., Wang, S., Tang, J., and Liu, H. 2017. Fake news detection on social media: A data mining perspective. ACM SIGKDD Explorations Newsletter, **19**(1), 22–36.

Socher, R., Perelygin, A., Wu, J. Y., Chuang, J., Manning, C. D., Ng, A. Y., and Potts, C. 2013. Recursive deep models for semantic compositionality over a sentiment treebank. Pages 1631–1642 of: *Proceedings of the Conference on Empirical Methods in Natural Language Processing.* EMNLP '13. Seattle: Association for Computational Linguistics.

Stede, M., and Schneider, J. 2018. Argumentation mining. *Synthesis Lectures on Human Language Technologies*, **11**(2), 1–191.

Stone, P. J., Dunphy, D. C., Smith, M. S., and Ogilvie, D. M. 1966. *The general inquirer: A computer approach to content analysis.* Cambridge: MIT Press.

Stoyanov, V., and Cardie, C. 2008. Topic identification for fine-grained opinion analysis. Pages 817–824 of: *Proceedings of the 22nd International Conference on*

Computational Linguistics. COLING '08. Manchester: The COLING 2008 Organizing Committee.

Strapparava, C., and Mihalcea, R. 2007 (June). SemEval-2007 Task 14: Affective text. Pages 70–74 of: *Proceedings of the Fourth International Workshop on Semantic Evaluations*. SemEval '17. Vancouver: Association for Computational Linguistics.

Struß, J. M., Siegel, M., Ruppenhofer, J., Wiegand, M., and Klenner, M. 2019. Overview of GermEval Task 2, 2019 shared task on the identification of offensive language. *Proceedings of the 15th Conference on Natural Language Processing*. KONVENS'2019. Erlangen: KONVENS 2019 Organizers.

Tai, K. S., Socher, R., and Manning, C. D. 2015. Improved semantic representations from tree-structured long short-term memory networks. Pages 1556–1566 of: *Proceedings of the 53rd Annual Meeting of the Association for Computational Linguistics and the 7th International Joint Conference on Natural Language Processing*. ACL-IJCNLP '15. Beijing: Association for Computational Linguistics.

Tang, D., Wei, F., Yang, N., Zhou, M., Liu, T., and Qin, B. 2014. Learning sentiment-specific word embedding for Twitter sentiment classification. Pages 1555–1565 of: *Proceedings of the 52nd Annual Meeting of the Association for Computational Linguistics*. ACL '14. Baltimore: Association for Computational Linguistics.

Tang, D., Qin, B., Feng, X., and Liu, T. 2016. Effective LSTMs for target-dependent sentiment classification. Pages 3298–3307 of: *Proceedings of the 26th International Conference on Computational Linguistics*. COLING '16. Osaka: The COLING 2016 Organizing Committee.

Thet, T. T., Na, J.-C., and Khoo, C. S. G. 2010. Aspect-based sentiment analysis of movie reviews on discussion boards. *Journal of Information Science*, **36**(6), 823–848.

Thorne, J., and Vlachos, A. 2018. Automated fact checking: Task formulations, methods and future directions. Pages 3346–3359 of: *Proceedings of the 27th International Conference on Computational Linguistics*. COLING '18. Santa Fe: Association for Computational Linguistics.

Thorne, J., Vlachos, A., Cocarascu, O., Christodoulopoulos, C., and Mittal, A. 2018. The fact extraction and verification (FEVER) shared task. Pages 1–9 of: *Proceedings of the First Workshop on Fact Extraction and VERification*. FEVER'2018. Brussels: Association for Computational Linguistics.

Tumasjan, A., Sprenger, T., Sandner, P., and Welpe, I. 2010. Predicting elections with Twitter: What 140 characters reveal about political sentiment. Pages 178–185 of: *Proceedings of the Fourth International Conference on Weblogs and Social Media*. ICWSM '10. Washington: AAAI Press.

Turney, P. D. 2002. Thumbs up or thumbs down?: Semantic orientation applied to unsupervised classification of reviews. Pages 417–424 of: *Proceedings of the Annual Meeting of the Association for Computational Linguistics*. New Orleans: Association for Computational Linguistics. ACL '02. Philadelphia: Association for Computational Linguistics.

Van Hee, C., Lefever, E., and Hoste, V. 2018. SemEval-2018 Task 3: Irony detection in English tweets. Pages 39–50 of: *Proceedings of the 12th International Workshop on Semantic Evaluation*. SemEval '18.

Vidgen, B., and Derczynski, L. 2020. Directions in abusive language training data: Garbage in, garbage out. PLOS ONE, **15**(12), 1–32.

Vosoughi, S., Roy, D., and Aral, S. 2018. The spread of true and false news online. *Science*, **359**(6380), 1146–1151.

Wang, X., Liu, Y., Sun, C., Wang, B., and Wang, X. 2015. Predicting polarities of tweets by composing word embeddings with long short-term memory. Pages 1343–1353 of: *Proceedings of the 53rd Annual Meeting of the Association for Computational Linguistics and the 7th International Joint Conference on Natural Language Processing*. ACL-IJCNLP '15. Beijing: Association for Computational Linguistics.

Wiebe, J., Wilson, T., Bruce, R., Bell, M., and Martin, M. 2004. Learning subjective language. *Computational Linguistics Journal*, **30**(3), 277–308.

Wiebe, J., Wilson, T., and Cardie, C. 2005. Annotating expressions of opinions and emotions in language. *Language Resources and Evaluation*, **39**(2–3), 165–210.

Wilson, T., Wiebe, J., and Hoffmann, P. 2005. Recognizing contextual polarity in phrase-level sentiment analysis. Pages 347–354 of: *Proceedings of the Conference on Human Language Technology and Empirical Methods in Natural Language Processing*. HLT-EMNLP '05. Vancouver: Association for Computational Linguistics.

Wilson, T., Wiebe, J., and Hoffmann, P. 2009. Articles: Recognizing contextual polarity: An exploration of features for phrase-level sentiment analysis. *Computational Linguistics*, **35**(3), 399–433.

Zampieri, M., Malmasi, S., Nakov, P., Rosenthal, S., Farra, N., and Kumar, R. 2019. SemEval-2019 Task 6: Identifying and categorizing offensive language in social media (OffensEval). Pages 75–86 of: *Proceedings of the 13th International Workshop on Semantic Evaluation*. SemEval '19. Minneapolis: Association for Computational Linguistics.

Zampieri, M., Nakov, P., Rosenthal, S., Atanasova, P., Karadzhov, G., Mubarak, H., Derczynski, L., Pitenis, Z., and Çöltekin, Ç. 2020. SemEval-2020 Task 12: Multilingual offensive language identification in social media. *Proceedings of the International Workshop on Semantic Evaluation*. SemEval '20. Stroudsburg: Association for Computational Linguistics.

Zhu, X., Guo, H., Mohammad, S. M., and Kiritchenko, S. 2014. An empirical study on the effect of negation words on sentiment. Pages 304–313 of: *Proceedings of the Annual Meeting of the Association for Computational Linguistics*. ACL '14. Baltimore: Association for Computational Linguistics.

PART II

Social Impact

Section Editors: Ivar Vermeulen and Camiel Beukeboom

4

Perspectives in a Social Context: The Role of Communication

Ivar Vermeulen and Camiel Beukeboom

When we, as empirical communication scientists, think about how to conceptualize a Perspective Web, the first approach that comes to mind is to investigate how perspectives manifest themselves. How do the perspectives that people base their ideas on, that organizations try to understand or influence, and that are engrained in institutions become observable? To communication scientists, the answer to this question is obvious: through communication.

Within the realm of "communication," a first place to look for perspectives is of course in messages being transferred. We can see perspectives in all messages great and small: in conversations with friends, in ads produced by marketeers, and in rules and regulations formalized by governments. But the communication process entails much more: it includes the sender of a message, who may have a particular perspective on a particular subject and a particular motivation to express it. It also includes the recipient, who may read a message and see his or her perspective confirmed or very much disputed. It includes the communication channel and its (often implicit) rules of engagement, so that users of one channel (e.g., social media) may have very different ideas on appropriate language and subject matter than users of another channel. It certainly includes the social and cultural context, where perspectives may differ between countries, subcultures, and social groups – between people at work and people in private. And finally, it includes effects – the changes that occur when all these factors come together: senders who strategically adapt their messages, recipients who adjust previously held ideas, social dynamics that create joy and anger, and that may unify or divide (see also Chapter 2).

Communication science is the scientific discipline that studies communication processes, as well as the messages being exchanged, their production, dissemination, selection, experience, and effects. Although this may seem like an almost impossibly broad domain to study (i.e., what societal phenomenon does not include communication?), it is conveniently constrained by a relatively limited set of theories that serve as a common framework for scholars to explain the phenomena they observe. And although communication scientists are also known as "prolific customers" of the "warehouse of theories" (usually of the linguistic, psychological, or sociological persuasion), they tend to rely on a fixed set of common explanations for a wide, and ever increasing, range of phenomena. Another thing that binds communication scientists is that they are usually in pursuit of societal goals: they want to lay bare implicit meaning (e.g., stereotypes) in messages, reduce negative effects of media use (e.g., cyberbullying, or misinformation), empower the use of media for the common good (e.g., reduce climate change, disparities, polarization), and promote pro-social and/or healthy behavior (e.g., volunteering, exercising). You will encounter all these elements in the three chapters that follow – written by communication scholars.

A distinction that is often made within communication science to fence off subdomains (e.g., for research groups, master tracks, or conference themes) are the spheres, or "levels," where communication processes take place. On the micro level, or individual sphere, communication occurs between individuals, in individual messages, or in individual media users. Think about the study of texts, of interpersonal verbal interactions, or of the feelings people experience when they watch a sad movie. On the meso level, or organizational sphere, the communication phenomena studied usually involve social structures such as groups or organizations: How do our intergroup biases affect how we interpret messages from individuals belonging to that group? How can the organizational identity of a source affect message selection in audiences? Finally, on the macro level, or public sphere, the communication is studied against the background of society and its institutions. How do media dynamics influence democratic processes? To what extent do social media platforms decrease or increase trust in government? The three contributions in the current section of this book discuss the notion of perspectives and the Perspective Web at these three levels: micro (individual), meso (group), and macro (public) communication. The fact that all authors are communication scholars becomes visible also in the theories and concepts they apply to elucidate phenomena, and in the problematic societal issues they address and aim to ameliorate.

In Chapter 5, Kobie Van Krieken and José Sanders explore perspectives in written discourse at a genuinely micro-level approach. Using a real-world example, they specifically focus on the role of viewpoints in narratives, and, show how, through this fairly explicit way of exposing the reader to different perspectives, readers' interpretations and attitudes may be affected.

In Chapter 6, Rachel Neo explores the meso level of communication, focusing on the currently most-studied way of message dissemination:

computer-mediated communication. By reviewing a number of leading theories that explain the production, selection, and reception of messages, she points at the different elements that the Perspective Web should consider in order to help users better interpret the messages they receive.

Finally, in Chapter 7, Hong Vu looks at the macro level of communication, aiming to answer the question of what the transparency, potentially brought about by the Perspective Web, will add to the current online media landscape, which is dominated by technological giants such as Google and Facebook. He combines a review of traditional communication theory and current empirical findings to discuss critical issues such as information algorithms and filter bubbles.

Together, the three chapters paint a broad picture of communication, and of all the communication processes to which the Perspective Web may potentially relate. The authors show that the amount of implicit information disseminated in messages, by senders and recipients, and embedded in channels, interpersonal relations, roles, organizations, and culture, can be vast and intimidating. Yet, they also outline the theoretical concepts and empirical regularities that may help start to unravel this massive and entangled heap of implicit information, and to – step by step – make the provenance, intentions, and underlying assumptions of the information we are exposed to more transparent.

5

The Micro Level: Linguistic Perspective in Written Discourse

Kobie van Krieken and José Sanders

5.1 Perspective in Language and Discourse

A multitude of concepts is covered by the word perspective.[1] In language alone, perspective may refer to ideas as different as subjective language, narrative perspective in literary fiction, and the reader's stance in the comprehension process. Apparently, perspective is a very fruitful concept that structures a range of discourse phenomena, and that is closely related to the concept of subjectivity. Speakers – as well as writers – express in their discourse where their perspective is located, in other words, from which point of view they are speaking. At first thought, it may seem that a monological discourse has one fixed perspective, that is, the writer's, whereas a dialogue or polylogue has as many perspectives as speakers. However, any natural discourse, either spoken or written, may show that this is not the case at all. This can be illustrated with an example from a natural journalistic discourse.[2] Context: in a corporate magazine of the Dutch national fire arms, a firefighter reports about a recent intervention. The primary target readership is people working for this organization. Example 5.1 provides the introductory lead of the article.

Example 5.1

1. P1 injury Intermezzo ice cream parlour, Heuvelring in Tilburg.
2. Commanding officer Pieter Jan Fonken was the first to receive this report on 22 July.

[1] This chapter is in part based on elements of Sanders (1994) and Sanders and Van Krieken (2018b).

[2] The Dutch article "Ik leef met jullie mee, veel sterkte" [I feel for you, best of luck] by Jildou Visser from the magazine Brand & Brandweer [Fire & Fire arms] in the series Onder de helm [Under the Helmet], available on https://issuu.com/sduuitgeversbrandbrandweer/docs/bb201710 (retrieved June 14, 2021 Brand & Brandweer 2017, nr. 10, p.31).

3. "We responded and drove to the fairground, without me having any idea of what had happened."
4. When he hears en route that a child is trapped under a fairground attraction, it goes quiet in the fire engine.
5. Fonken: "'Oh shit' was my first thought.
6. 'I sympathise with you guys, lots of luck', added the operator.
7. Operators never say that.
8. Afterwards, it turned out he'd seen images from a surveillance camera in the vicinity."

In this fragment, the speaking voice (in this case, journalistic narrator) is – necessarily – in the first place talking from her[3] own perspective, although she does not mention herself explicitly as "I." She chooses what to tell, and how, dependent on what she wants to achieve with the text. For instance, it is not explained what a P1 injury is because it is already known to the audience. But it is clear that within the text she is taking various other perspectives. In the first sentence, an objective viewpoint is taken, merely indicating the scene and the type of incident. In the second sentence, the journalist introduces the person who participates in the intervention and experiences it from within: she tells what Pieter Jan Fonken learned. In the fourth sentence, another actor enters the scene, but still the journalist takes Fonken's perspective. This impression is caused by the fact that the narrator says what Fonken hears instead of the operator telling him that a child is trapped; this indicates that the journalist takes Fonken's position to describe the scene. However, the text goes on to say that it goes quiet in the fire engine. Here it seems that the journalist is moving a little bit away from Fonken and positions herself in the middle of a group of firefighters in the engine.

In the next few sentences, we see that Fonken and the operator both have their own perspective: they refer to themselves in first person (my and I, respectively) and to the other in second person (you). They also have their own views on the situation, which are expressed by subjective language: statements such as *Oh shit* (sentence 5), and exclamations such as *lots of luck* (sentence 6). Fonken's and the operator's expressions may be exemplary for their professional positions. As a fire fighter, Fonken is experienced in dealing with such situations and knows from the unusual fact that the operator wishes them luck (sentence 7) that a severe situation will have to be met. The operator, as intermediate, is not immediately involved; from this distant perspective, wishing the fire fighters good luck is an expression of empathy. Fonken does not know

[3] The story was written by a female journalist.

at that point in time that the operator has seen images of the critical situation (sentence 8), which cause him or her to express sympathy (sentence 7). Readers of this text may tend to empathize with Fonken because the journalist started the text with his perspective, which primes him as main character of the text. Another reason, less connected with the text than with the readers' own perspectives, is that most readers of this corporate magazine can imagine the firefighter's situation more readily than the operator's context. Also, the factor that readers may be attracted toward the firefighter's heroic role more than toward the operator's facilitating role may direct the readers' empathy choice (Sanders and Van Krieken, 2018a). This in turn will influence the interpretation of the text: for instance, readers who strongly empathize and identify with Fonken may feel suspense because they share his worry about the scene he will arrive at shortly.

The analysis of this short fragment shows that many different kinds of perspectives can be discerned in written discourse and that these perspectives may influence the reading process in terms of perspective-taking, empathy, and identification. Ultimately, this may affect readers' attitudes, beliefs, knowledge, intentions, and behavior (e.g., De Graaf et al., 2012; Shen et al., 2014; see also De Graaf et al., 2016). This chapter cannot describe and explain all these aspects of discourse perspective exhaustively. Instead, it will concentrate on the way a speaker presents the perspective of persons, specifically in narrative texts, and the functions this perspective-taking has.

5.2 A Text-Linguistic Approach to Perspective in Narrative

In this chapter, we will take a text-linguistic approach to perspective in discourse. The starting point in this approach is that discourse is a strategic means of communication, with which the interactive subjects have particular objectives in exchanging their discourses. Their objectives are dependent on context: an advertisement wants to convince the audience to buy a product or service, while a story in a corporate magazine wants to reinforce employees' and other stakeholders' commitment to the company. A fundamental difference regarding perspective is that between discourse representing a story world and discourse representing a state of affairs. In the latter, indicated as exposition, objects and elements are described in their complex cohesion (Graesser et al., 2003). This type of text is abstract and lacks any internal time-lapse. Representing a mental image of such a text requires the conceptualization of a network of facts and connections with a rather general validity (Georgakopoulou and Goutsos, 2000), as in (constructed) example 5.2:

Example 5.2

(2) The Netherlands is divided into twenty-five fire brigade regions. In many regions, the firefighters are volunteers, but there are also professional corps in the large cities. In addition to fighting fires, the fire brigade is also tasked with providing information about fire safety.

Hence, an exposition is explicitly not about consecutive events and experiences of persons in a specific time frame at a specific place. A story, by contrast, is precisely that: a series of temporally connected events placed in a specific context that are perceived and experienced by an experiencing subject. Representing a story requires the conceptualization of a narrative microworld of temporally connected events and situations. Both expository and narrative discourse can have a broad range of functions, for example: to inform, to convince, or to divert, but they do so in different ways. For narrative discourse to be effective in achieving its goals, the perspective from which the story world is construed plays a crucial role.

Perspective is a complex concept that, on a micro level, refers to the way in which, by linguistic means, events in the story world are told based on the sensory perceptions and/or physical, psychological, or ideological viewpoint of the specific person in the narrative (henceforth: character). At a macro level, perspective refers to both the way in which the character is portrayed (from the "inside" or the "outside") and the way in which others refer to the character. For example, several studies have assessed whether the audience identifies more with a character referred to in the first person (an "I" character), in the second person (a "you" character), or in the third person (a "he" or "she" character). Results showed that the second person evokes more identification (Andeweg et al., 2013) and more emotions (Brunyé et al., 2011) than the first person. However, in practice, stories written in the second person are rare; the vast majority of stories feature first and/or third person characters. Both these types of stories are effective in "transporting" the audience into the story world (see e.g. Banerjee and Greene, 2012), but, in general, readers seem to be able to identify more strongly with I characters than with he or she characters (Segal et al., 1997). Within the context of health communication, stories are also often found to be more persuasive when told in the first person rather than in the third person (see De Graaf et al., 2016).

However, this does not mean that a story told in the third person cannot be persuasive or evoke identification. Identification with a he or she character is influenced by linguistic choices. For example, a narrator can refer to a character with a name and/or full noun (Pieter Jan Fonken/the man) or with a personal pronoun (she). A personal pronoun creates proximity between character and reader, while a noun creates distance. Persistent reference to a character with a pronoun rather than a noun therefore facilitates readers to identify with this character (Van Krieken and Sanders, 2017).

There are various ways of expressing perspective within stories in the first person and stories in the third person. As readers, we may not only be given access to the visual and auditive perspective of a character but also to their speech, thoughts, emotions, and actions. Narrators have a wide range of linguistic elements at their disposal to express each of these viewpoint dimensions (Van Krieken et al., 2017). These expressions of perspective form the capillaries of a story and determine, to a large extent, how readers will process the story, that is to say: how close a character is emotionally and from which point of view the reader imagines the events (Van Krieken, 2018). In literary theory, these perspective phenomena are often discussed in terms of focalization.

5.3 Focalization

To describe the phenomenon that the events in a narrative are being perceived and imagined from a particular subjective perspective, the concept of focalization is used, distinguishing between internal and external focalization (Bal, 1985; Genette, 1980). In the case of external focalization, the narrator is the perceiving entity, such as in *After his shift, Fonken returns home*. Here, the character is perceived and described from the outside by the narrator. In the case of internal focalization, a character is a perceiving entity, such as in *When he hears en route that a child is trapped under a fairground attraction, it goes quiet in the fire engine*; or in *At first, Fonken doesn't see the child.* Without literally representing the character's inner thoughts or experiences, implicit references to this inner world may be made by expressions of sensory perceptions (He *saw* the trapped child), emotions (He *was concerned* for the trapped child), modal constructions (He *had to cut* through the cables around the trapped child) and cognitive actions (He *decided to cut* the trapped child free).

Subtle expressions of internal focalization often go unnoticed, but are in fact indicators of implicit viewpoint that guide readers into processing the narrative events and situations from the inside perspective of the character (Sanders and

Redeker, 1993; Van Krieken et al., 2017). Although external and internal focalization can alternate within a story, often one of the two forms of perspective predominates. The choice between using predominantly external focalization or predominantly internal focalization is functional and evokes a specific kind of processing in readers that has consequences for the way in which they learn from the knowledge enclosed in narrative discourse.

5.4 Demonstration versus Invasion

A story that shows various actions and situations mainly from the perspective of an external focalizer demonstrates the story events (Sanders, 2017). From such a demonstration, readers can experience the events in the story world along with the characters, thereby deriving experiential knowledge. In a cognitive sense, simulating the events (whether factual or fictional) in the discourse allows the audience to co-experience them and to absorb the experiential knowledge enclosed in the story, which can be added to their own knowledge. A demonstrative story thus grants its audience the freedom to draw their own conclusions.

Internal focalization offers an alternative and more direct way of transferring knowledge by penetrating the realm of thought of a character,in the story. By way of such an invasion, the audience can immediately adopt the character's reasoning. This implies that an invasive story with internal focalization articulates the message in an explicitly subjective way, particularly by verbalizing the perceptions and evaluations of the focalizing character. By contrast, in a demonstrative story, more cognitive steps are required to make the knowledge enclosed in the narrative discourse explicit. In case of an unwelcome message, the explicit articulation of an invasive narrative can evoke more resistance, such as in *He was sure of it. Merely stopping smoking had saved his life*! On the other hand, such an invasive story is conceptually more straightforward than a demonstrative story, in which the reasoning is more implicit, such as in *He stopped smoking. Gradually, his coughing decreased.* For instance, health messages in demonstrative narratives could be too complex for target audiences with a lower level of literacy (De Graaf et al., 2017). In other words, which type of story will have a beneficial effect in which context will depend on the target group. The effectiveness of the type of story can also be presumed to depend on the objective of the story: a demonstrative story can be particularly effective when the objective is to inform, persuade, and activate readers, while an invasive story can be particularly effective when the objective is to involve readers affectively.

To illustrate the difference between demonstration and invasion, Table 5.1 compares demonstrative and invasive versions of one story fragment from the Brand&Brandweer text. The demonstrative version has an external focalizer and describes the events "from the outside." This can be done in both the third person (Column 1) and the first person (Column 2). An invasive story gives the reader access to the inner experiences of the character. This internally focalizing character can be a third-person (Column 3) or first-person (Column 4) character. Hence, a story told in the third person can be either demonstrative (Column 1) or invasive (Column 3), depending on the degree to which access is gained to the perceptions and evaluations of the he or she character. In a story told in the first person, access to the inner world of the I character is by definition available. For that reason, an I story is in principle invasive. However, that does not necessarily have to be the case: if access to the inner world of the I character is not realized, we have a demonstrative story (Column 2). Many strategic stories used by organizations fall into the category of an invasive story told in the third person (Column 3). In addition, we often see the inner world of this character through direct I quotations.

This example shows that the demonstration is characterized by descriptions of the situation and the actions performed by the character. In the invasive fragments, the character is presented not only as one who acts but also as a character who perceives and evaluates. In addition to perceptions and evaluations, stories also represent statements and thoughts. This is done using indirect, direct, and free indirect quotations. We will go on to discuss the diverse categories of quotations in more detail.

5.5 Quotations

A narrator can use various ways of quoting, that is: representing the speech and thoughts of characters. In this chapter, we follow the classification by Semino and Short (2004), whose frequently used categorization model for the representation of speech and thoughts was based on a large-scale study into narrative and nonnarrative texts. In the first place, a narrator can indicate that a character speaks or thinks without specifying what exactly is being said or thought. In such a case, we speak of the representation of a speech act, in the words of Semino and Short (2004), a narrator's report of speech act, as in *He talked for hours* or *He talked about the fire brigade for hours*. With the representation of a speech act, the narrator presents a character as a speaking and thinking person. Such a representation is strongly diegetic in nature; in other words, the narrator tells about speech or thoughts of a character. The narrator can also choose to let characters speak for themselves. This kind of representation is strongly mimetic

Table 5.1 *Example of text fragment in demonstrative and invasive styles*

Column 1: Demonstration Third person	Column 2: Demonstration First person	Column 3: Invasion Third person	Column 4: Invasion First person
At that point, the little boy was trapped under the ride car. Fonken and the crew of the first fire truck stabilized the car, then freed and removed it. The little boy's arm was stuck. Fonken removed the covers from the bottom of the attraction, giving him a better view of the arm that was stuck. He and the commanding officers of the second fire truck and the other emergency vehicle then started by cutting through the railing with the blowtorch. Time was running out. Their combined efforts got the little boy free.	At that point, the little boy was trapped under the ride car. I and the crew of the first fire truck stabilized the car, then freed and removed it. The little boy's arm was stuck. I removed the covers from the bottom of the attraction, giving me a better view of the arm that was stuck. I and the commanding officers of the second fire truck and the other emergency vehicle then started by cutting through the railing with the blowtorch. Time was running out. Our combined efforts got the little boy free.	At that point, Fonken didn't know what the little boy was caught in. He and the crew of the first fire truck stabilized the car, then freed and removed it. He saw that the little boy's arm was stuck. Fonken removed the covers from the bottom of the attraction, and only then saw exactly how the arm was stuck. He was afraid it was going to go wrong. Where should he start? He and the commanding officers of the second fire truck and the other emergency vehicle decided to start by cutting through the railing with the blowtorch to get a better view. They felt time pressing. Only by working together, and quickly, did they succeed!	At that point, I didn't know what the little boy was caught in. I and the crew of the first fire truck stabilized the car, then freed and removed it. I saw that the little boy's arm was stuck. I removed the covers from the bottom of the attraction, and only then did I see exactly how the arm was stuck. "This is never going to work," I thought. "Where on earth am I supposed to start?" I and the commanding officers of the second fire truck and the other emergency vehicle decided to start by cutting through the railing with the blowtorch to get a better view. We could feel time running out. Only by working together, and quickly, did we succeed!

in nature; in other words, the narrator shows the content of a character's speech and thoughts. The mimetic representation of speech and thoughts can fulfil a number of functions. In the first place, speaking and thinking allows characters to be more lifelike: they are given a human voice. For example, in the fire brigade story, the main character says: "'Oh shit' was my first thought." This makes it easier for the reader to vividly imagine the character. Secondly, and particularly in journalistic and corporate contexts, quotations may fulfil a demonstrative function by showing that the story is true and based on information from reliable sources (see Van Krieken and Sanders, 2016; Van Krieken et al., 2016). For example: "This is never going to work, I thought. It looked really complicated," said Fonken. A quotation like this demonstrates the veracity of the story, and that is important because the effectiveness of a corporate story depends entirely on the extent to which the readers regard the story as being an authentic one. The content of the speech and thoughts of characters can be represented in various forms of speech. We will discuss the three major forms: direct, indirect, and free indirect speech.

5.6 Direct Speech

Direct speech is a literal representation of a character's utterances. Placing the utterances in inverted commas and having the character speak in the I form is characteristic of direct speech. Hence, in the case of a story told in the third person, there is a shift to the first person in references to the character: *[After his shift, Fonken returns home.] "When I got home, I told my story again. I don't usually talk about operations, but this one was special, and I'll never forget it."* Direct speech has a dramatizing function: the narrator lets a character speak in their own words, that's to say in their own style and register (Thompson, 1996). A direct quotation can be seen, in this regard, as a mini I story embedded in the large story.[4]

5.7 Indirect Speech

In indirect speech, the words of the character are represented less literally, and more paraphrased by the narrator. That means that although the character may be responsible for the content of what is said, the narrator is responsible for the form. The result is that the character's voice and that of the narrator are intertwined to some extent (Sanders, 2010). A characteristic of indirect speech is that the character's quotation is embedded syntactically, generally in a dependent clause beginning with "that" *(He said that...)* or "if" *(He asked if ...)*.

[4] If there are no inverted commas around a direct quotation and no introductory sentence or phrase, we speak of free direct speech: At home, I ...

Since the literalness of the representation of speech and thoughts is not of primary importance in indirect speech, there is less room for being expressive, and this type of speech is less lively and dramatic in nature than direct speech (Toolan, 2006).

The degree of paraphrasing in indirect speech is variable. A narrator can stick closely to the character's words and represent them quite literally *(He said that he never usually talked about work, but that this operation had been special and he would never forget it)*, or, conversely, they can summarize a long explanation by the character into a single sentence *(He said that this operation had been memorable)*. In addition, a narrator can choose to represent a small stretch of the character's words in inverted commas *(He said that this operation had been "special")*. A partial quotation like this in indirect speech is typically an emotional, controversial, or witty statement by the character (Vis et al., 2015). By placing the words in inverted commas, the narrator can enliven the quotation while simultaneously observing a distance from the content of the statement (Wieland, 2010).

5.8 Free Indirect Speech

A speech form which combines the characteristics of both direct and indirect speech is free indirect speech. Just as in indirect speech, voices are intertwined in free indirect speech (Sanders, 2010): the voice of the character sounds through the voice of the narrator. Free indirect speech is, however, syntactically not embedded as indirect speech is but follows the same word order as direct speech. The major difference with direct speech is that no switch from the third to the first person takes place in free indirect speech, such as in *He thought back. No, he would truly never forget this operation.* Note that this can also occur in the first person, but it is less noticeable if it does: *I thought back. No, I would truly never forget this operation.* Free indirect speech is a form of speech and thought representation that was first described for literary fiction, but it is in fact found in all other genres in which it is functional to represent the inner world of a character.[5]

In essence, the representation of speech and thought is a potentially recursive process. Characters who tell a story within the narrative can quote themselves or other persons in their story. Thus, within direct speech, a character can quote themselves or someone else, and this, again, can take the form of a direct, indirect, or free indirect representation. For example, in the opening

[5] Free indirect thought is also known as stream of consciousness; see, for example, Cohn, 1978.

sentences of the fire brigade story, a direct quote of the main character, Pieter Jan Fonken, contains two embedded cases of direct quotation: first, he represents his own thoughts and then allows the operator to speak. (Fonken: "'Oh shit' was my first thought. 'I sympathise with you guys, lots of luck', added the operator.")

5.9 Multiple Perspectives

That subjective perspective is intrinsically recursive requires readers to construct complex mental representations of the discourse. Keeping track of the embedding of perspectives within perspectives is a challenging task, specifically in cases where this embedding is achieved implicitly without attributing the information to a particular source. For instance, in the fire brigade's story, the abovementioned direct quote is surrounded by implicit attributions:

Example 5.3

> 4. When he hears en route that a child is trapped under a fairground attraction, it goes quiet in the fire engine. [5. Fonken: "'Oh shit' was my first thought. 6. 'I sympathise with you guys, lots of luck', added the operator.] 7. Operators never say that. 8. Afterwards, it turned out he had seen images from a surveillance camera in the vicinity."

In (4), Fonken is reported to hear about the critical situation; from their schematic world knowledge, readers may infer that this information was retrieved by telecommunication with some operator. Indeed, an operator is subsequently introduced in sentence (6) and readers may infer backwards that the report in (4) can positively be attributed to this source. In yet another way, the information in sentence (8) is also attributed to this source. Here, Fonken reports what the operator at that moment in the story knew without saying it but revealed at a later moment (presumably, the gravity of the situation). Note that it is not made explicit whether this information was delivered directly by the operator to Fonken. It is also possible that an unknown intermediate source transferred it to Fonken: either way, readers will have to trust that there is indeed a source for this information and that it is trustworthy. An interesting case, in addition, is Fonken's claim in (7): Operators never say that. This is a general statement of the expository type, which may serve to underscore the rarity of the operator's wish. Readers will have to accept that Fonken, as a firefighter, is a trustworthy source for this general claim.

The analysis of even this small discourse illustrates the complexity of embedded perspectives within perspectives, particularly in terms of the attribution of information to explicit or implicit sources. The inclusion of multiple perspectives in written discourse creates multiple layers of information that cumulatively steer readers' understanding, interpretation, and evaluation of the discourse. In this process, it may not always be possible for readers to correctly link the individual information layers to the corresponding sources, resulting in a blurring of sources and information. This multiplicity is even multiplied in stories that are created and transferred in social media. Micro-stories of societally relevant incidents are transferred, transformed, critiqued, and ultimately backgrounded or erased in favor of more personalized and localized interpretations by a process of stancetaking and rescripting of the initial incident (Georgakopoulou, 2014). Multiplied implicit and possibly skewed perspectives of unrevealed sources in the retelling of incidents will increasingly determine how these incidents are made sense of. Distinguishing between explicitly marked and implicit, merely inferable source perspectives is a task of great relevance. The Perspective Web (or Web5.0) may help to increase source transparency by disentangling the various layers of information and providing readers, so to speak, with a legend to identify and encode each layer. Ultimately, this could result in the interpretation of narrative discourse through an incremental process based on the integration of clearly distinguished information layers, each linked to an individual source, rather than an undirected process of moving through a hazy network of intertwined layers and perspectives. As discussed in this chapter, perspectives in discourse are established by a wide range of micro-level linguistic phenomena. Taking these phenomena into account in the creation of the Perspective Web may therefore be pivotal in its potential to enhance the transparency of online information.

References

Andeweg, S., Hendrix, R., Hoff, V. V. T., and Hoop, H. D. 2013. En dan vermoord je hem… Over de invloed van grammaticale persoon op identificatie [And then you kill him. On the influence of grammatical person on identification]. *Neerlandistiek*, **2013**, 1–19.

Bal, M. 1985. *Narratology: Introduction to the theory of narrative*. Toronto: University of Toronto Press.

Banerjee, S. C., and Greene, K. 2012. Role of transportation in the persuasion process: Cognitive and affective responses to antidrug narratives. *Journal of Health Communication*, **17**(5), 564–581.

Brunyé, T. T., Ditman, T., Mahoney, C. R., and Taylor, H. A. 2011. Better you than I: Perspectives and emotion simulation during narrative comprehension. *Journal of Cognitive Psychology*, **23**(5), 659–666.

Cohn, D. 1978. *Transparent minds: Narrative modes for presenting consciousness in fiction*. Princeton, NJ: Princeton University Press.

De Graaf, A., Hoeken, H., Sanders, J., and Beentjes, H. 2012. Identification as a mechanism of narrative persuasion. *Communication Research*, **39**(6), 802–823.

De Graaf, A., Sanders, J., and Hoeken, H. 2016. Characteristics of narrative interventions and health effects: A review of the content, form, and context of narratives in health-related narrative persuasion research. *Review of Communication Research*, **4**, 88–131.

De Graaf, A., Van den Putte, B., Nguyen, M.-H., Zebregs, S., Lammers, J., and Neijens, P. 2017. The effectiveness of narrative versus informational smoking education on smoking beliefs, attitudes and intentions of low-educated adolescents. *Psychology & Health*, **32**(7), 810–825.

Genette, G. 1980. *Narrative discourse*. Oxford: Blackwell.

Georgakopoulou, A. 2014. Small stories transposition and social media: A micro-perspective on the "Greek crisis." *Discourse & Society*, **25**(4), 519–539.

Georgakopoulou, A. and Goutsos, D. 2000. Revisiting discourse boundaries: The narrative and non-narrative modes. *Text-Interdisciplinary Journal for the Study of Discourse*, **20**(1), 63–82.

Graesser, A. C., McNamara, D. S., and Louwerse, M. M. 2003. What do readers need to learn in order to process coherence relations in narrative and expository text? Pages 82–98 of: Sweet, A. P. and Snow, C. E. (eds.), *Rethinking reading comprehension*. New York: Guilford.

Sanders, J. 1994. Perspective in narrative discourse. Ph.D. thesis, Tilburg University.

Sanders, J. 2010. Intertwined voices: Journalists' modes of representing source information in journalistic subgenres. *English Text Construction*, **3**(2), 226–249.

Sanders, J. 2017. We hebben een verhaal nodig. Inaugural lecture, Radboud University Nijmegen.

Sanders, J. and Redeker, G. 1993. Linguistic perspective in short news stories. *Poetics*, **22**(1), 69–87.

Sanders, J. and Van Krieken, K. 2018a. Exploring narrative structure and hero enactment in brand stories. *Frontiers in Psychology*, **9**, 1645.

Sanders, J. and Van Krieken, K. 2018b. Narrative analysis [narratieve analyse]. Pages 199–238 of: Karreman, J. and Van Enschot, R. (eds), *Text analysis: Methods and applications [Tekstanalyse: Methoden en toepassingen]*. Assen: Van Gorcum.

Segal, E. M., Miller, G., Hosenfeld, C., Mendelsohn, A., Russell, W., Julian, J., and Delphonse, J. 1997. Person and tense in narrative interpretation. *Discourse Processes*, **24**(2–3), 271–307.

Semino, E. and Short, M. 2004. *Corpus stylistics: Speech, writing and thought presentation in a corpus of English writing*. London: Routledge.

Shen, F., Ahern, L., and Baker, M. 2014. Stories that count: Influence of news narratives on issue attitudes. *Journalism & Mass Communication Quarterly*, **91**(1), 98–117.

Thompson, G. 1996. Voices in the text: Discourse perspectives on language reports. *Applied Linguistics*, **17**(4), 501–530.

Toolan, M. 2006. Speech and thought: Representation of. Pages 698–710 of: Brown, E. K. and Anderson, A. (eds), *Encyclopedia of language & linguistics*. Boston, MA: Elsevier.

Van Krieken, K. 2018. Ambiguous perspective in narrative discourse: Effects of viewpoint markers and verb tense on readers' interpretation of represented perceptions. *Discourse Processes*, **55**(8), 771–786.
Van Krieken, K. and Sanders, J. 2016. Diachronic changes in forms and functions of reported discourse in news narratives. *Journal of Pragmatics*, **91**, 45–59.
Van Krieken, K. and Sanders, J. 2017. Engaging doctors and depressed patients: Effects of referential viewpoint and role similarity in health narratives. *International Journal of Communication*, **11**, 1–18.
Van Krieken, K., Sanders, J., and Hoeken, H. 2016. Blended viewpoints, mediated witnesses: A cognitive linguistic approach to news narratives. Pages 145–168 of: Dancygier, B., Lu, W.-I., and Verhagen, A. (eds), *Viewpoint and the fabric of meaning: Form and use of viewpoint tools across languages and modalities*. Berlin: Mouton de Gruyter.
Van Krieken, K., Hoeken, H., and Sanders, J. 2017. Evoking and measuring identification with narrative characters: A linguistic cues framework. *Frontiers in Psychology*, **8**, 1–16.
Vis, K., Sanders, J., and Spooren, W. 2015. Quoted discourse in Dutch news narratives. Pages 152–172 of: Levie, S., Hoeken, H., and Lüthy, C. (eds), *Texts, transmissions, receptions: Modern approaches to narrative texts.*
Wieland, N. 2010. Context sensitivity and indirect reports. *Philosophy and Phenomenological Research*, **81**(1), 40–48.

5.10 Appendix: I Sympathise with You Guys, Lots of Luck

Source: Brand&Brandweer, SdU publishers: October 2017
Author: Jildou Visser
Translation: Radboud In'to Languages, Radboud University, Nijmegen 2017

P1 injury Intermezzo ice cream parlour, Heuvelring in Tilburg. Commanding officer Pieter Jan Fonken was the first to receive this report on 22 July. "We responded and drove to the fairground, without me having any idea of what had happened." When he hears en route that a child is trapped under a fairground attraction, it goes quiet in the fire engine. Fonken: "'Oh shit' was my first thought. 'I sympathise with you guys, lots of luck', added the operator. Operators never say that. Afterwards, it turned out he'd seen images from a surveillance camera in the vicinity."

The firefighters are met at the fairground by a traffic controller. "I could see from the look on the traffic controller's face, and the way he waved us through, that the incident was a serious one. After that, everything went so quickly. Fifty metres further, and we'd arrived. There were lots of people, screaming in panic." At first, Fonken doesn't see the child. Bystanders point the victim out, lying jammed under a ride car about two metres in the air. "It didn't look good." The ride, for small children, consists of a row of cars, linked together, that ride slowly upwards and downwards. "The little boy had fallen out of a ride car and landed underneath it. Bystanders were already lifting the ride car, so the full weight of it wasn't on his stomach any more. The little boy was conscious and very calm. He didn't scream once during the whole operation, but the look in his eyes spoke volumes. His eyes were screaming for help. Now and then he closed them. When we started talking to him, he opened them and looked at us with that expression."

At that point, the firefighters don't know how exactly the little boy is trapped. The crew of the first TS truck stabilizes the ride car, then disconnects and removes it. Only then can they see that the little boy's arm is jammed in the slit of the mechanism that moves the cars forward. "It wasn't until we removed the covers from underneath the ride, that we could see properly how his little arm was stuck. This is never going to work, I thought. It looked really complicated," said Fonken. He and the commanding officers of the second fire truck and the other emergency vehicle decide to start by cutting through the railing with the blowtorch to get a better view. At that point, a bystander starts shouting at the fire fighters. "A woman walked towards one of my crew just as he was about to cut through the railing. He looked at me. Neither of us knew who she was. She might have been a concerned parent, but when she started pulling at my crew member, it was clear. Luckily, the police intervened quickly," Fonken explains. "Then there's a commotion a bit further away, when a second person turns on the police. You're just taken aback for a moment, that you're trying to save a child and people are turning on you like that. After that, all attention went back to the operation and the child, while the police dealt with the fight further along."

6

The Meso Level: Perspectives in a Social Context

Rachel Neo

6.1 The Meso Level

In the field of communication, meso-level communication research largely focuses on how cues within one's social environment influence perceptual and behavioral outcomes. This stands in contrast to the micro and macro levels of understanding and analyzing communication phenomena. The micro level focuses on the analysis of viewpoints represented in written discourse, whereas the macro level focuses on how larger cultural and technological forces shape communication processes.

People do not exist in isolated silos. Rather, they are inherently social creatures who care about the opinions of others (Noelle-Neumann, 1974). With the proliferation of Web 2.0 platforms, it is now common for people to be exposed to online information from a myriad of sources and messages (Walther and Jang, 2012). Such online information and cues are ubiquitously present in online social environments and play instrumental roles in shaping attitudes, judgments, and behaviors. This chapter draws on theories of computer-mediated communication and media effects to explain how people produce and react to messages in social environments at the meso level.

6.2 Theories of Computer-Mediated Communication and New Media

Communication typically consists of exchanges between message senders and recipients. In online environments, people are message "prosumers" – they both produce and send messages, as well as receive and react to messages (Yamamoto et al., 2020). The first part of this chapter discusses relevant computer mediated communication (CMC) and new media theories from communication science and social psychology that identify and examine the social factors influencing message production and reception in online environments.

6.2.1 The Social Identity Model of Deindividuation Effects (SIDE)

The Social Identity Model of Deindividuation Effects (SIDE) argues that the relative lack of nonverbal cues in online environments causes individual identities to become less salient (Lea and Spears, 1992; Reicher et al., 1995). People will therefore tend to view both themselves and others in terms of group membership characteristics, not individual differences. In turn, they will conform to norms of the online group that they identify with (Lea and Spears, 1992; Reicher et al., 1995).

Many studies have used SIDE as a theoretical framework to examine how message receivers respond to other people in online environments. These studies generally tend to be experiments that examine the effects of visual anonymity and group identification on response valence. For instance, a study showed that students who participated in online synchronous text-based discussions with anonymous conversation partners and were primed to believe that they were part of a social group were most likely to conform to perceived group norms (Spears et al., 1990). Interestingly, although the SIDE model explains how message receivers orient themselves to others in terms of group identities, some scholars contend that SIDE can help to account for interpersonal relationship dynamics in contexts such as online romantic relationship formation (Lea and Spears, 1995). Interpersonal cues such as nonverbal behaviors and perceived physical attractiveness are not necessarily precursors to romantic relationship development. Rather, in online anonymous environments, romantic relationships can develop because of perceptions of a shared sense of social identity between message senders and receivers (Sanders, 1997). Furthermore, the tenets of the SIDE theory have received support even in the context of collated online content such as threads of user comments. Specifically, people are likely to form content attitudes that are in line with the valence of these comments (Chung, 2018; Walther et al., 2010b). For example, people are likely to form negative attitudes about online content that receives negative user comments.

6.2.2 Social Information Processing Theory

Unlike the SIDE model, which focuses on the deindividuating effects of online environments, the social information processing (SIP) theory focuses on how people form interpersonal relationships with others in online versus face-to-face communication contexts (Walther, 1992). SIP acknowledges the relative lack of nonverbal cues in online environments vis-à-vis face-to-face communication but posits that people are able to form online interpersonal relationships that are of similar quality to offline relationships over time. Specifically, the SIP explains how message senders use CMC platforms to build rapport and how recipients in turn form impressions of these message senders.

The SIP assumes that message senders want to reduce uncertainty about their conversation partners and will attempt to build intimacy and rapport with others regardless of communication platform (Walther, 1992). Although certain types of CMC platforms might lack nonverbal cues, message senders will adapt their communication styles to suit the capabilities of these platforms (Walther, 1992). Research has shown that message senders using real-time CMC chat software will use certain verbal cues instead of nonverbal cues to convey unfriendliness toward their conversation partners when discussing controversial issues (Walther et al., 2005).

Furthermore, message senders will strategically use visuals such as photos instead of text to convey impressions of themselves to their conversation partners in CMC settings (Tanis and Postmes, 2003). More recently, research has examined how unobtrusive cues aid in impression formation within online environments. Some studies have shown that people still rely more on interactive online communication than unobtrusive cues such as social media profile photographs to reduce uncertainty about the followers within their network (Antheunis et al., 2010). However, the type of unobtrusive cue can affect perceptions of partner attractiveness. Participants who saw nonhuman avatars exhibited more uncertainty about their partner's attractiveness than those exposed to photographs (Westerman et al., 2015).

In addition, the SIP contends that message receivers will take a longer time to form impressions of their conversations partners in CMC contexts than in face-to-face communication contexts (Walther, 1992). When people engage in verbal communication on synchronous CMC platforms, their conversation partners are unable to visualize and decode their verbal reactions at the same rate as a face-to-face conversation. However, other scholars have advocated for a more nuanced approach to understanding how CMC platforms aid in relationship formation. Newer types of CMC platforms are able to support multiple modes of verbal and nonverbal communication (Westerman et al., 2008). Specifically, these scholars have argued that multi-modal CMC communication platforms are more conducive for impression development because such platforms convey more information cues than CMC platforms that allow for only textual communication. Indeed, research on the SIP has shown that online photos and videos, which convey more information than mere text, help expedite impression formation on the part of message receivers (Westerman et al., 2008).

6.2.3 Hyperpersonal Model

The hyperpersonal model of CMC (Walther, 1996) goes one step further than the SIP by contending that CMC-based relationship development and intimacy surpasses that of offline communication. In particular, the hyperpersonal model explains how sender goals affect message production in online environments. People engage in strategic self-presentation because they deliberately want to influence others' opinions of them. Walther (1996) argues that text-based CMC facilitates selective self-presentation. People could take advantage of the lack of nonverbal cues on these platforms to consciously construct and convey a desired online persona to their conversation partners. Such platforms also allow message senders to communicate with their conversation partners without displaying undesirable nonverbal communication behaviors such as yawning or a lack of eye contact (Walther, 2011).

Notably, the hyperpersonal model identifies a few strategies that people use to strategically foster rapport with their conversation partners. First, research has shown that strangers use more intimate self-disclosures and utterances in CMC to foster rapport with their conversations partners than in offline interactions (Joinson, 2001; Tidwell and Walther, 2002). Second, people deliberately use verbal convergence strategies in CMC contexts to create a sense of affinity with their conversation partners, and vice versa (Walther et al., 2010a). Third, people take advantage of the editing capabilities of asynchronous CMC platforms to carefully craft messages with the goal of impressing socially desired targets (Walther, 2011). Notably, people even go to the rather extreme extent of deliberately misrepresenting their physical attributes such as age, weight, or height on online dating sites so as to attract potential romantic partners (Hall et al., 2010; Toma et al., 2008).

In addition, the hyperpersonal model contends that message receivers can react to cues from message senders in their online social environment. In a process akin to behavioral confirmation, message receivers will first idealize strategically presented information from message senders. They will then provide feedback that causes the senders to modify their identities in ways that align with the receivers' idealized impressions of them (Gonzales and Hancock, 2008). In an experiment, participants first received a message instructing them to answer questions in either an introverted or extroverted manner. They were further told to post their responses to these questions on CMC platforms. Next, half of the participants received feedback affirming their introversion/extroversion, whereas the other half did not. Participants who received personality-affirming feedback were most likely to give self-reports of their introversion/extroversion that aligned with the feedback (Walther et al., 2011).

6.2.4 Warranting Theory

The term "warrant" refers to perceptions of authenticity associated with online information describing a person or entity (Walther and Parks, 2002). Many scholars and previous research have argued that online environments enable people to hide certain flaws of their offline selves and present distorted images of themselves to others (Donath, 1999). Though people might learn about other people through online interactions (Ellison et al., 2006), such online information may not necessarily be valid or credible (Walther and Parks, 2002).

As such, one key tenet of the warranting theory contends that message senders are less likely to engage in deceptive self-presentation if receivers are able to corroborate the sender's claims with accurate, real-life information from their peers or offline contacts (Walther, 2011). In order to persuade potential romantic partners of their trustworthiness, people on online dating sites often pair textual descriptions of their activities, for example, rock climbing, with photos of them engaging in such activities (Ellison et al., 2006). It would arguably be rather difficult for people to fabricate photos of themselves participating in strenuous physical activities such as rock climbing (Donath, 1999). Another study showed that people were less likely to lie about their physical attributes on dating sites if they knew that peers from their offline social circle used such sites (Toma et al., 2008). Also, online chat users whose identities were linked with offline information were less likely to engage in deceptive self-presentation about their preferences and physical attributes on such platforms (Warkentin et al., 2010).

In addition, the warranting theory explains how message receivers evaluate the credibility of online content. Message receivers are likely to be more confident about the veracity of online content that is easily traced or linked to a legitimate offline persona (Sharabi and Dykstra-DeVette, 2019; Walther, 2011). Online information that is "immune to manipulation by the person to whom it refers" (Walther and Parks, 2002, p.552) is more likely to be perceived as credible. Quite a few studies have shown support for this aspect of the warranting theory. People are more likely to trust other-generated statements such as comments made by one's peers more than self-generated statements such as claims made by social media profile owners (Walther et al., 2009). Qualitative content analyses of email messages to potential romantic partners have shown that message receivers proactively took steps to verify their prospective partner's online identities by requesting their social media handles or to "friend" them on social media (Sharabi and Dykstra-DeVette, 2019). However, other studies have shown that people's evaluations of a target may contradict other-generated evidence, such as sociometric cues, that aligns with self-generated claims. For example, research showed a curvilinear relationship between the number of Facebook friends and ratings of profile owner popularity (Tong et al., 2008). Although the number of Facebook friends, a sociometric cue, was consistent with self-generated claims about popularity, people were likely to think that those with too many Facebook friends were artificially padding their popularity levels by increasing their friend count beyond realistic levels.

6.2.5 The MAIN Model

The MAIN model derives itself from information-processing theories such as the heuristic-systematic processing model to explain how four broad types of cues in online social environments – Modality (M), Agency (A), Interactivity (I), and Navigability (N) cues – trigger heuristics (mental shortcuts) in message receivers that influence credibility perceptions (Sundar, 2008).

Modality Cues

The term "modality" refers to the mode (e.g., text, aural, audiovisual) through which information is conveyed online. With digital convergence, online platforms allow for the multi-modal presentation of information. Various mode(s) of presenting online information can trigger cognitive heuristics (Sundar, 2008).

For instance, Sundar (2008) has argued that audiovisual modes of information depict scenarios in a more realistic manner than textual modes of information. Furthermore, textual information has more nuances and requires more cognitive effort to parse out than audiovisual information. The realism heuristic nonetheless posits that people will trust audiovisual information more than textual information. Although message receivers might be cognizant of the fact that senders can deceitfully manipulate or doctor audiovisual information, they nonetheless will still perceive audiovisual information to be a more authentic depiction of the real world than the latter (Reeves and Nass, 1996; Sundar, 2008). However, presenting information in multi-modal formats might impede cognitive processing capabilities (Sundar, 2000, 2008). Online platforms often overwhelm people with a plethora of cues. As such, the distraction heuristic posits that multi-modal information can cause cognitive overload, potentially resulting in negative credibility evaluations of online content (Sundar, 2000). For example, if people are overstimulated by the multimodal presentation of Web content, they might be less likely to trust such content (Sundar, 2000).

Agency Cues

Agency cues pertain to the sources of online information. These cues can take the form of proprietor-generated sources (such as websites or a news organization), user-generated sources (such as a comment left by a friend on one's Facebook profile), or aggregated information sources that collectively indicate users' preferences in the form of online rating scores (Walther and Jang, 2012). In particular, aggregated user ratings trigger what is known as the bandwagon heuristic ("If others think something is good, it must be good!"). Many studies have shown that online ratings are extremely persuasive cues that influence a myriad of outcomes such as product attitudes and purchase intentions (Sundar et al., 2008), newsworthiness (Sundar et al., 2007), and news article selection (Messing and Westwood, 2014). However, other studies have outlined boundary conditions under which online ratings have limited persuasive efficacy. For instance, the effects of online rating scores on information trustworthiness depend on rating volume. People are less likely to use rating scores to evaluate information trustworthiness when rating volume is low (Flanagin and Metzger, 2013). Also, rating scores that disaffirm strongly held issue beliefs can cause boomerang effects. When people distrust rating scores, there will be a negative association between rating scores and information trustworthiness such that favorable scores cause information distrust and vice versa (Neo, 2018).

Furthermore, online sources can influence credibility perceptions via the social presence heuristic by creating the perception that one is communicating with a social actor (Sundar, 2008). The "computers as social actors" perspective argues that people anthropomorphize computers by ascribing human traits and characteristics when interacting with them. The more online platforms foster a sense of social presence through human-like vocal or audiovisual features, the more users are likely to trust online information (Sundar, 2008). For example, research has shown that positive online book reviews read by multiple synthetic voices fostered a greater sense of social presence than book reviews read by one synthetic voice. In turn, social presence was positively associated with website credibility, personal opinions of the book, and public opinion toward the book (Lee and Nass, 2004).

Interactivity Cues

Although the concept of "interactivity" is widely regarded as a hallmark of online platforms, there is still no consensus as to what this term actually means.

Some scholars regard interactivity as the degree to which message senders and receivers can engage in two-way communication with one another via online platforms (Sundar, 2008). Yet others construe interactivity to mean the degree to which changes or modifications that people make to an online interface will be reflected in real time (Lee et al., 2006).

In terms of heuristics, interactive tabs or pull-down menus can trigger the choice heuristic (Wang, 1998). When people think that they have the freedom to decide what types of online content to look at, they are likely to have favorable evaluations of the Web interface. However, scholars caution that the choice heuristic can trigger negative evaluations of online interfaces if content is presented in a disorganized manner (Goodhue, 1995). Furthermore, certain types of interactive features can make the control heuristic cognitively accessible (Eveland et al., 2002; Schwartz, 2000). When people believe that online features enable them to control the types of online information they see (Eveland et al., 2002), they will develop favorable impressions of information credibility.

Navigability Cues

Navigability refers to an online interface's ability to enable people to move from one online page to another. An interface's design and organizational structure can trigger heuristics associated with navigability. The hyperlinks on websites enable people to search for information in a nonlinear manner (Eveland et al., 2002). For example, websites that list hyperlinks can prime the browsing heuristic and prompt the message receiver to explore various facets of the website (Khan and Locatis, 1998). When people believe that they have the ability to browse through a website and its related content, they are likely to form positive impressions of content authenticity and overall website credibility. By contrast, hyperlinks that are seamlessly intertwined with online content can trigger the elaboration heuristic. These types of hyperlinks could potentially increase cognitive elaboration by helping individuals to make cognitive connections between various site offerings (Eveland et al., 2002).

6.3 Traditional Theories of Media Effects

Before the advent and proliferation of online platforms, scholars coined media effects theories to explain how people were influenced by information disseminated by traditional media platforms such as the television or print newspapers. Although the social Web comes with different affordances than traditional platforms, these media effects theories are still highly germane in this digital age, and complement the CMC theories outlined above in primarily explaining how people react to, or process information on the social Web.

6.3.1 Selective Exposure and Avoidance

The selective exposure hypothesis posits that people exhibit clear preferences toward information that aligns with their beliefs (Frey, 1986). Festinger's theory on cognitive dissonance can help explain the phenomenon of selective exposure (Festinger, 1957). People experience psychological discomfort when they are exposed to information that is inconsistent with their beliefs. In order to minimize cognitive dissonance, people are more likely to seek out and consume information from sources that they agree with (Frey, 1986).

In recent years, the selective exposure hypothesis has received considerable attention from political communication scholars. The political media landscape has evolved rapidly – from a few channels offered by traditional media outlets to an abundance of traditional and online political information sources (Sunstein, 2001). As mentioned above in the MAIN model, online platforms and collaborative filtering technologies allow users to exercise control over the types of content they consume. Some scholars have argued that people will take advantage of the control afforded by these online environments to deliberately expose themselves exclusively to belief-affirming political content (Sunstein, 2001; Valentino et al., 2009). Thus far, studies have demonstrated robust evidence for the selective exposure phenomenon. People generally prefer to consume pro-attitudinal online political content (e.g., Garrett, 2009), and spend more time reading pro-attitudinal than counter-attitudinal online information (Knobloch-Westerwick et al., 2015).

Furthermore, selective exposure has implications for political outcomes. Research has demonstrated a fairly strong positive effect of selective exposure on political participation (Knobloch-Westerwick and Johnson, 2014). The more people selectively expose themselves to pro-attitudinal information, the more confidence they will have in their political attitudes. Such increased confidence in political attitudes makes people more likely to take political action (Knobloch-Westerwick and Johnson, 2014). In addition, selective exposure to pro-attitudinal online news has a positive direct effect on affective polarization – a phenomenon characterized by favoritism toward the in-party and hostility toward the out-party (Garrett et al., 2014). Importantly, scholars have stressed the importance of decoupling selective exposure from selective avoidance, a practice in which people deliberately eschew counter-attitudinal content (Garrett, 2009). People might overwhelmingly prefer to consume pro-attitudinal political information, but that does not necessarily mean that they would avoid counter-attitudinal information (Garrett, 2009; Knobloch-Westerwick and Kleinman, 2012). There are situations in which people will deem it beneficial to consume counter-attitudinal information. First, people are likely to consume useful counter-attitudinal content over pro-attitudinal content lacking in utility (e.g., Freedman, 1965). Experiments have demonstrated that Republicans chose to consume counter-attitudinal political content containing useful information about the out-party candidate over pro-attitudinal political content when they realized that their candidate was going to lose the 2008 election (Knobloch-Westerwick and Kleinman, 2012). Second, people might deliberately expose themselves to counter-attitudinal political information in order to develop arguments against opposing political viewpoints (Garrett et al., 2013). Indeed, research has shown that anxiety impels counter-attitudinal political information consumption among people who want to rebut opposing political viewpoints (Valentino et al., 2009). The emotionally gratifying process of successfully counter-arguing against attitudinal dissonant political information helps people to reduce their anxiety (Valentino et al., 2009).

6.3.2 Spiral of Silence

The spiral of silence theory by Noelle-Neumann (1974) is one of the most "researched and controversial approaches to understanding public opinion" (Glynn et al., 1999, p. 203) and has been hailed as a public opinion theory with a "reasonable degree of [...] range, depth and coverage" (Neuwirth, 2000, p.140). According to the spiral of silence theory, people are social creatures who fear being socially isolated (Noelle-Neumann, 1974). They will thus scan the social environment for cues regarding the so-called climate of opinion and use their quasi-statistical sense to gauge the nature of other people's opinions on an issue (Scheufele and Moy, 2000). If individuals perceive that their stances on the issue are consonant with the majoritarian view or gaining in popularity, they will be more likely to publicly express their opinions on the issue. However, if they perceive their opinions to be in the minority, or losing ground, they will be more likely to fall silent on the issue (Noelle-Neumann, 1974).

Early CMC research on the spiral of silence theory showed that CMC environments were more effective than face-to-face communication at increasing willingness to express one's honest opinions in the face of a hostile opinion climate (Ho and McLeod, 2008). Scholars argued that the visual anonymity afforded by CMC platforms made people more disinhibited and consequently more likely to espouse minority viewpoints (Ho and McLeod, 2008). However, more recent research carried out with multi-modal online platforms such as social media have shown support for the spiral of silence (Gearhart and Zhang, 2014; Stoycheff, 2016). For instance, perceived hostile opinion climate was negatively associated with one's willingness to speak out on Facebook. This negative relationship was strongest among individuals who perceived that the government was monitoring their social media activities and felt that such surveillance was justified (Stoycheff, 2016). In addition, willingness to self-censor and perceived opinion incongruence with the general public were negatively associated with willingness to speak out about issues related to the bullying of LGBT individuals on social media (Gearhart and Zhang, 2014).

6.3.3 Parasocial Interactions

The term "parasocial interactions" refers to the relationships and connections that audiences develop with a celebrity through mediated platforms, and typically measures the extent to which people perceive celebrities to be their friends (Horton and Richard Wohl, 1956). Unlike offline or mediated interpersonal relationships, these relationships tend to be one-sided in nature because the celebrities rarely, if ever, reciprocate affection from an audience member. People often consider parasocial relationships with celebrities to be an important aspect of their social environment (Schramm and Wirth, 2010).

With the rise of popular online video streaming sites such as YouTube, scholars have begun to examine how people develop parasocial relationships with YouTube or social-media celebrities (Berryman and Kavka, 2017; Choi and Behm-Morawitz, 2017; Ferchaud et al., 2018). These YouTube celebrities have built their career almost entirely through creating a cohesive brand image of themselves through their video content (Ferchaud et al., 2018). In particular, Lange (2007) defines a YouTube celebrity as one who utilizes their interactions to "influence the discourse, goals, and activities" of the platform (p. 5), often by engaging in self-expression and portraying aspects of their self-identity (Berryman and Kavka, 2017).

Scholars have argued that YouTube personalities function as message producers who deliberately try to foster parasocial connections with their online audiences by creating the false impression that their connections with audience members are reciprocal (Labrecque, 2014). A quantitative content analysis of the most-subscribed YouTube channels showed that these YouTube celebrities attempt to create parasocial interactions with their fans through positive, negative, and neutral self-disclosures on vlogs. The amount of self-disclosure was positively associated with perceived YouTuber authenticity. In addition, videos featuring females were perceived as more realistic as compared to videos with male YouTubers. Also, the direction of a YouTuber's gaze affects perceptions of authenticity. Coders were more likely to perceive videos featuring on-camera faces as being more authentic than videos without any on-camera actors or actresses (Ferchaud et al., 2018).

In terms of audience reception toward YouTube celebrities, qualitative studies have indicated that the parasocial interaction strategies outlined above help to cultivate rapport with audience members (Berryman and Kavka, 2017). Furthermore, research has shown that quantity of popularity cues such as subscriber count positively predicts depth of parasocial interactions with YouTube beauty vloggers and video watching enjoyment. Furthermore, there is a positive association between video subscriber count and product purchase intentions. Taken together, these findings indicate the bandwagon heuristic at play. People use bandwagon cues such as subscriber counts to determine the degree to which they should bond with YouTube beauty vloggers or buy their products (Rasmussen, 2018). Another study showed that watching videos of YouTube celebrities had a positive effect on perceived quality of parasocial relationships via perceptions of social and physical attraction toward these celebrities (Kurtin et al., 2018).

6.4 Conclusion

This chapter draws upon selected CMC and media effects theories to explain how social factors drive peoples' production and reception of online messages. Some theories, such as the social information processing theory, the hyperpersonal model, and the warranting theory, explain how socially motivated reasons work in tandem with online affordances to influence both message production and reception (Walther, 2011). Other theories such as selective attention (Frey, 1986) and the MAIN model focus more on how message receivers process cues and messages within their online social environment (Sundar, 2008).

Pre-existing dispositions such as attitudes, values, or motivations work together with social Web affordances to shape the (a) type(s) of information used in message production and (b) how information is received by message recipients. In some instances, there is a symbiotic relationship between the information used in message production and reception via the social Web. For example, the hyperpersonal model states that message producers are motivated to engage in strategic impression management, and will take advantage of the anonymity afforded by online platforms to craft and disseminate information that casts them in a favorable light. In turn, message receivers are motivated to use available information to form impressions of the message sender. The anonymity afforded by online platforms is likely to cause a message recipient to form idealized impressions of the message senders in line with the information provided by the message sender. The parasocial identity model also alludes to that by arguing that people idealize and form relationships with the social media influencers whom they follow on social media. When people form parasocial relationships with social media influencers, they are likely to readily accept the very information put forth by such influencers.

However, there can be discrepancies in information used in message production and reception via the social Web. Notably, message receivers might perceive things differently by selectively honing in on other aspects of the message instead of the information elements intended by the message sender. For example, health campaign practitioners are motivated to elicit prosocial behaviors among their intended audience. They might craft health campaign messages hoping that such messages will go viral and generate positive online word of mouth. However, the message might backfire if the audience reacts to the message in ways that the practitioners did not foresee, for example, deeming certain aspects of the message culturally offensive and an affront to their values, and indicating their displeasure in the form of bandwagon cues such as "down votes" or user-generated content such as negative comments. Other intended audience members might become aware of the online backlash surrounding the message, and likewise form perceptions of the message that align with the negative comments or unfavorable downvotes.

In addition, the capabilities and message features within online social environments, as well as individual-level predispositions, can influence receivers' reactions to messages. For example, receivers are likely to believe claims about a message sender or target if they perceive that such information cannot easily be falsified (Walther et al., 2009). People will trust Facebook statuses made by a sender or target if these claims align with Facebook comments from the message sender's peers, and vice versa (Walther et al., 2009). Although interface features in online social environments can be very persuasive, there are boundary conditions under which these cues operate. For example, bandwagon cues such as rating scores are only likely to trigger the bandwagon heuristic when rating volume is high (Flanagin and Metzger, 2013). Furthermore, message receivers are likely to reject bandwagon cues that disaffirm strongly held issue beliefs (Neo, 2018). In addition, message receivers will also use certain online interfaces or message cues such as comment valence to form perceptions about the opinion climate. Such perceptions of the opinion climate can either stymie or encourage frank opinion expression (Stoycheff, 2016).

Furthermore, the control afforded by online platforms enables message receivers to choose what they want to see (e.g., Sundar, 2008; Sunstein, 2001). The more people believe these platforms give them control over what they read or watch, the more they will have favorable evaluations of the online platform

(Eveland et al., 2002). Importantly, message receivers use online platforms to selectively expose themselves to pro-attitudinal online information (Garrett, 2009). They are especially likely to trust such online information if it aligns with their attitudes (Metzger et al., 2010).

All in all, online interface cues and messages can potentially be integral parts of a Perspective Web that sheds light on how information spreads, the sources behind information creation and dissemination, the parties or clusters of groups most likely to be receptive to such information, and how certain types of information are distorted, for example. However, regardless of the Perspective Web's potential, some communication/media effects theories explicitly state that people are prone to using pre-existing values or attitudes to process information, for example, selective attention theory. Other theories such as the spiral of silence also state that perceptions of social reality trump objective representations of data and reality. As such, despite the potential the Perspective Web holds, people might reject objective online information or data about information veracity/sources. Consequently, there will still be inherent challenges in changing stubborn minds, or convincing people to reject falsehoods and embrace accurate information.

References

Antheunis, M. L., Valkenburg, P. M., and Peter, J. 2010. Getting acquainted through social network sites: Testing a model of online uncertainty reduction and social attraction. *Computers in Human Behavior*, **26**, 100–109.

Berryman, R. and Kavka, M. 2017. I guess a lot of people see me as a big sister or a friend: The role of intimacy in the celebrification of beauty vloggers. *Journal of Gender Studies*, **26**, 307–320.

Choi, G. Y. and Behm-Morawitz, E. 2017. Giving a new makeover to STEAM: Establishing YouTube beauty gurus as digital literacy educators through messages and effects on viewers. *Computers in Human Behavior*, **73**, 80–91.

Chung, J. E. 2018. Peer influence of online comments in newspapers: Applying social norms and the social identification model of deindividuation effects (SIDE). *Social Science Computer Review*. Advance online publication.

Donath, J. 1999. Identity and deception in the virtual community. Pages 29–59 of: Smith, M. A. and Kollock, P. (eds), *Communities in cyberspace*. New York: Routledge.

Ellison, N. B., Heino, R. D., and Gibbs, J. L. 2006. Managing impressions online: Self-presentation processes in the online dating environment. *Journal of Computer-Mediated Communication*, **11**, 415–441.

Eveland, Jr, P. W. and Dunwoody, S. 2002. An investigation of elaboration and selective scanning as mediators of learning from the Web versus print. *Journal of Broadcasting & Electronic Media*, **46**, 34–53.

Ferchaud, A., Grzeslo, J., Orme, S., and LaGroue, J. 2018. Parasocial attributes and YouTube personalities: Exploring content trends across the most subscribed YouTube channels. *Computers in Human Behavior*, **80**, 88–96.

Festinger, L. 1957. *A theory of cognitive dissonance*. Stanford, CA: Stanford University Press.

Flanagin, A. J. and Metzger, M. J. 2013. Trusting expert versus user-generated ratings online: The role of information volume, valence, and consumer characteristics. *Computers in Human Behavior*, **29**, 1626–1634.

Freedman, J. L. 1965. Confidence, utility, and selective exposure: A partial replication. *Journal of Personality and Social Psychology*, **2**, 778–780.

Frey, D. 1986. Recent research on selective exposure to information. *Advances in Experimental Social Psychology*, **19**, 41–80.

Garrett, R. K. 2009. Echo chambers online? Politically motivated selective exposure among Internet news users. *Journal of Computer-Mediated Communication*, **14**, 265–285.

Garrett, R. K., Carnahan, D., and Lynch, E. K. 2013. A turn toward avoidance? Selective exposure to online political information. *Political Behaviour*, **2004–2008**(35), 113–134.

Garrett, R. K., Gvirsman, S. D., Johnson, B. K., Tsfati, Y., Neo, R., and Dal, A. 2014. Implications of pro-and counterattitudinal information exposure for affective polarization. *Human Communication Research*, **40**, 309–332.

Gearhart, S. and Zhang, W. 2014. Gay bullying and online opinion expression: Testing spiral of silence in the social media environment. *Social Science Computer Review*, **32**, 18–36.

Glynn, C. J., Herbst, S., O'keefe, G. J., and Shapiro, R. Y. 1999. *Public opinion*. 1st ed. Boulder, CO: Westview Press.

Gonzales, A. L. and Hancock, J. T. 2008. Identity shift in computer-mediated environments. *Media Psychology*, **11**, 167–185.

Goodhue, D. L. 1995. Understanding user evaluations of information systems. *Management Science*, **41**, 1827–1844.

Hall, J. A., Park, N., Song, H., and Cody, M. J. 2010. Strategic misrepresentation in online dating: The effects of gender, self-monitoring, and personality traits. *Journal of Social and Personal Relations*, **27**, 117–135.

Ho, S. S. and McLeod, D. M. 2008. Social-psychological influences on opinion expression in face-to-face and computer-mediated communication. *Communication Research*, **35**(2), 190–207.

Horton, D. and Richard Wohl, R. 1956. Mass communication and para-social interaction: Observations on intimacy at a distance. *Psychiatry*, **19**, 215–229.

Joinson, A. N. 2001. Self-disclosure in computer-mediated communication: The role of self-awareness and visual anonymity. *European Journal of Social Psychology*, **31**, 177–192.

Khan, K. and Locatis, C. 1998. Searching through cyberspace: The effects of link display and link density on information retrieval from hypertext on the World Wide Web. *Journal of the American Society for Information Science*, **49**, 176–182.

Knobloch-Westerwick, S. and Johnson, B. K. 2014. Selective exposure for better or worse: Its mediating role for online news' impact on political participation. *Journal of Computer-Mediated Communication*, **19**, 184–196.

Knobloch-Westerwick, S. and Kleinman, S. B. 2012. Preelection selective exposure: Confirmation bias versus informational utility. *Communication Research*, **39**, 170–193.

Knobloch-Westerwick, S., Mothes, C., Johnson, B. K., Westerwick, A., and Donsbach, W. 2015. Political online information searching in Germany and the United States: Confirmation bias, source credibility, and attitude impacts. *Journal of Communication*, **65**, 489–511.

Kurtin, K. S., O'Brien, N., Roy, D., and Dam, L. 2018. The development of parasocial interaction relationships on YouTube. *The Journal of Social Media in Society*, **7**(1), 233–252.

Labrecque, L. I. 2014. Fostering consumer-brand relationships in social media environments: The role of parasocial interaction. *Journal of Interactive Marketing*, **28**, 134–148.

Lange, P. G. 2007. Commenting on comments: Investigating responses to antagonism on YouTube. Pages 163–190 of: *Society for applied anthropology conference*, vol. 31. Tampa, FL: Society for Applied Anthropology.

Lea, M. and Spears, R. 1992. Paralanguage and social perception in computer-mediated communication. *Journal of Organizational Computing*, **2**, 321–341.

Lea, M. and Spears, R. 1995. Love at first byte? Building personal relationships over computer networks. Page 197–233 of: Wood, J. T. and Duck, S. (eds), *Understudied relationships: Off the beaten track.* Thousand Oaks, CA: Sage.

Lee, K. M. and Nass, C. 2004. The multiple source effect and synthesized speech: Doubly-disembodied language as a conceptual framework. *Human Communication Research*, **30**, 182–207.

Lee, K. M., Park, N., and Jin, S. 2006. Narrative and interactivity in computer games. Pages 259–274 of: Vorderer, P. and Bryant, J. (eds), *Playing video games*. Mahwah, NJ: Erlbaum.

Messing, S. and Westwood, S. J. 2014. Selective exposure in the age of social media: Endorsements trump partisan source affiliation when selecting news online. *Communication Research*, **41**, 1042–1063.

Metzger, M. J., Flanagin, A. J., and Medders, R. B. 2010. Social and heuristic approaches to credibility evaluation online. *Journal of Communication*, **60**, 413–439.

Neo, R. L. 2018. The limits of online consensus effects: A social affirmation theory of how aggregate online rating scores influence trust in factual corrections. *Communication Research*. Advance online publication.

Neuwirth, L. 2000. Testing the spiral of silence model: The case of Mexico. *International Journal of Public Opinion Research*, **12**, 138–159.

Noelle-Neumann, E. 1974. The spiral of silence: A theory of public opinion. *Journal of Communication*, **24**, 43–51.

Rasmussen, L. 2018. Parasocial interaction in the digital age: An examination of relationship building and the effectiveness of YouTube celebrities. *The Journal of Social Media in Society*, **7**(1), 280–294.

Reeves, B. and Nass, C. I. 1996. *The media equation: How people treat computers, television, and new media like real people and places*. New York: Cambridge University Press.

Reicher, S. D., Spears, R., and Postmes, T. 1995. A social identity model of deindividuation phenomena. *European Review of Social Psychology*, **6**, 161–198.

Sanders, R. E. 1997. Find your partner and do-si-do: The formation of personal relationships between social beings. *Journal of Social and Personal Relationships*, **14**, 387–415.

Scheufele, D. and Moy, P. 2000. Twenty-five years of spiral of silence: A conceptual review and empirical outlook. *International Journal of Public Opinion Research*, **12**, 3–28.

Schramm, H. and Wirth, W. 2010. Testing a universal tool for measuring parasocial interactions across different situations and media: Findings from three studies. *Journal of Media Psychology: Theories, Methods, and Applications*, **22**, 26–36.

Schwartz, B. 2000. Self-determination: The tyranny of freedom. *American Psychologist*, **55**(1), 79–88.

Sharabi, L. L., and Dykstra-DeVette, T. A. 2019. From first email to first date: Strategies for initiating relationships in online dating. *Journal of Social and Personal Relationships*, **36**(11–12), 3389–3407.

Spears, R., Lea, M., and Lee, S. 1990. De-individuation and group polarization in computer-mediated communication. *British Journal of Social Psychology*, **29**, 121–134.

Stoycheff, E. 2016. Under surveillance: Examining Facebook's spiral of silence effects in the wake of NSA Internet monitoring. *Journalism & Mass Communication Quarterly*, **93**, 296–311.

Sundar, S. S. 2000. Multimedia effects on processing and perception of online news: A study of picture, audio, and video downloads. *Journalism & Mass Communication Quarterly*, **77**, 480–499.

Sundar, S. S. 2008. The MAIN model: A heuristic approach to understanding technology effects on credibility. Pages 73–100 of: Metzger, M. J. and Flanagin, A. J. (eds), *Digital media, youth, and credibility*. Cambridge, MA: The MIT Press, The John D. and Catherine T. MacArthur Foundation Series on Digital Media and Learning.

Sundar, S. S., Knobloch-Westerwick, S., and Hastall, M. R. 2007. News cues: Information scent and cognitive heuristics. *Journal of the American Society for Information Science and Technology*, **58**, 366–378.

Sundar, S. S., Oeldorf-Hirsch, A., and Xu, Q. 2008. The bandwagon effect of collaborative filtering technology. Pages 3453–3458 of: *Proceedings of CHI '08 Extended Abstracts on Human Factors in Computing Systems*. New York: Association for Computing Machinery.

Sunstein, C. R. 2001. *Republic.com*. Princeton, NJ: Princeton University Press.

Tanis, M. and Postmes, T. 2003. Social cues and impression formation in CMC. *Journal of Communication*, **53**, 676–693.

Tidwell, L. C. and Walther, J. B. 2002. Computer-mediated communication effects on disclosure, impressions, and interpersonal evaluations: Getting to know one another a bit at a time. *Human Communication Research*, **28**, 317–348.

Toma, C. L., Hancock, J. T., and Ellison, N. B. 2008. Separating fact from fiction: An examination of deceptive self-presentation in online dating profiles. *Personality and Social Psychology Bulletin*, **34**, 1023–1036.

Tong, S. T., Van Der Heide, B., Langwell, L., and Walther, J. B. 2008. Too much of a good thing? The relationship between number of friends and interpersonal impressions on Facebook. *Journal of computer-mediated communication*, **13**(3), 531–549.

Valentino, N. A., Banks, A. J., Hutchings, V. L., and Davis, A. K. 2009. Selective exposure in the Internet age: The interaction between anxiety and information utility. *Political Psychology*, **30**, 591–613.

Walther, J. B. 1992. Interpersonal effects in computer-mediated interaction: A relational perspective. *Communication Research*, **19**, 52–90.

Walther, J. B. 1996. Computer-mediated communication: Impersonal, interpersonal, and hyperpersonal interaction. *Communication Research*, **23**, 3–43.

Walther, J. B. 2011. Theories of computer-mediated communication and interpersonal relations. Pages 443–479 of: Knapp, M. L. and Daly, J. A. (eds), *The handbook of interpersonal communication*. Thousand Oaks, CA: Sage.

Walther, J. B. and Jang, J. W. 2012. Communication processes in participatory websites. *Journal of Computer-Mediated Communication*, **18**, 2–15.

Walther, J. B. and Parks, M. R. 2002. Cues filtered out, cues filtered in: Computer-mediated communication and relationships. Pages 529–563 of: Knapp, M. L. and Daly, J. A. (eds), *Handbook of interpersonal communication*. Thousand Oaks, CA: Sage.

Walther, J. B., Loh, T., and Granka, L. 2005. Let me count the ways: The interchange of verbal and nonverbal cues in computer-mediated and face-to-face affinity. *Journal of Language and Social Psychology*, **24–65**, 36.

Walther, J. B., Van Der Heide, B., Tong, T., S., Carr, C. T., and Atkin, C. K. 2010a. The effects of interpersonal goals on inadvertent intrapersonal influence in computer-mediated communication. *Human Communication Research*, **36**, 323–347.

Walther, J. B., DeAndrea, D., Kim, J., and Anthony, J. 2010b. The influence of online comments on perceptions of anti-marijuana public service announcements on YouTube. *Human Communication Research*, **36**, 469–492.

Walther, J. B., Van Der Heide, B., Hamel, L. M., and Shulman, H. C. 2009. Self-generated versus other-generated statements and impressions in computer-mediated communication: A test of warranting theory using Facebook. *Communication Research*, **36**(2), 229–253.

Walther, J. B., Liang, Y. J., DeAndrea, D. C., Tong, S. T., Carr, C. T., Spottswood, E. L., and Amichai-Hamburger, Y. 2011. The effect of feedback on identity shift in computer-mediated communication. *Media Psychology*, **14**(1), 1–26.

Wang, R. Y. 1998. A product perspective on total data quality management. *Communications of the ACM*, **41**(2), 58–66.

Warkentin, D., Woodworth, M., Hancock, J. T., and Cormier, N. 2010. Warrants and deception in computer-mediated communication. Pages 9–12 of: Inkpen, K. and Gutwin, C. (eds), *Proceedings of the 2010 ACM conference on computer supported cooperative work*. New York: ACM.

Westerman, D., Tamborini, R., and Bowman, N. D. 2015. The effects of static avatars on impression formation across different contexts on social networking sites. *Computers in Human Behavior*, **53**, 111–117.

Westerman, D. K., Van Der Heide, B., Klein, A. K., and Walther, J. B. 2008. How do people really seek information about others? Information seeking across Internet and traditional communication sources. *Journal of Computer-Mediated Communication*, **13**, 751–767.

Yamamoto, M., Nah, S., and Bae, S. Y. 2020. Social media prosumption and online political participation: An examination of online communication processes. *New Media & Society*, **22**, 1885–1902.

7

The Macro Level: Perspectives Embedded in Society, Culture, and Technology

Hong Vu

7.1 Introduction

In the digital age, the Internet and online entities like Google and Bing have provided and facilitated unprecedentedly vast amounts of information from a myriad of sources, offering users access to a range of opinions far broader than those found in a single newspaper or broadcast channel (Flaxman et al., 2016). Adequate representation of perspectives is essential for a pluralist media system as it encourages various population groups in the society to participate in the political process, which is important to the functioning of democracies (Möller et al., 2018). In principle, the Internet and its surrounding technologies are often expected to revive the public sphere, narrowing the temporal and spatial gaps between these groups to boost cross-culture exposure and increase interests in understanding others' viewpoints (Papacharissi, 2002). In reality, information abundance, brought about by advance digital technologies, has not always been a boon for everyone. Too much information overloads users (York, 2013). Confronting deviating perspectives overwhelms, confuses, and irritates audiences, causing some of them to retreat into their cocoons through avoiding information altogether (Bawden and Robinson, 2009; Park, 2019; Zhang et al., 2016). Many have sought ways to navigate in the sea of information, trying to either filter out information they do not want or go straight to the content they find interesting, useful, and agreeable to process in their limited amounts of time (Dylko et al., 2017). To aid that, modern-day digital platforms have increasingly employed social and technical mechanisms to limit users' exposure to attitude-challenging information through selectively customizing their content to their users' interests (Bakshy et al., 2015).

The primary goal of this chapter is to answer the question: In the context of the current mediascape, is it necessary to have a Perspective Web, where the perspectives of the sources of the information and knowledge that we encounter online shall be made transparent to users? To achieve that goal, this chapter begins by discussing such mechanisms in the context of the ubiquitous use of large online platforms such as Facebook and Google and their implications for the information landscape. It then focuses on revisiting a few related theoretical concepts such as selective exposure, homophily, echo chambers, and filter bubbles. This chapter also discusses the role of recommender systems and technical mechanisms in diffusing information and makes suggestions for technological solutions that might do a better job of compromising between creating a seamless online experience for users and informing users of what might really be happening in the world.

7.2 Technologies for Personalized Content

A healthy information environment means users can be exposed to content from diverse perspectives. Ideally, digital technologies and their incomparable ability should work to remove barriers that could help build a Perspective Web, allowing users to have the opportunity to see all information related to the content they encounter. However, so far, digital technological developments seem to have gone down another path. More than two decades since Google was founded in 1998, the world's largest search engine has been widely used across the world. "To google" something became the word of the year in 2002 and was officially included in the Oxford English Dictionary as an intransitive verb in 2006 (Heffernan, 2017), suggesting its global success. Not too many of us, however, have ever taken a look at what the tech giant says about itself: "Our mission is to organize the world's information and make it universally accessible and useful" (Google, 2019). All sounds benign. What is interesting here is the "organize" part. How does Google do it? What criteria does it base its algorithms on to "organize" and diffuse information to users?

PageRank has been one of the primary solutions Google employed since the early days to organize users' search results. According to Google (2019, p. 1), PageRank algorithms are designed to sort through billions of websites on the Internet "to find the most relevant, useful results in a fraction of a second" and present search results in a way that helps users find what they are looking for. Google is not alone in taking the liberty to "organize" information for Internet users. Almost every commercial site, including popular ones such as Yahoo, Bing, Facebook, Twitter, YouTube, Amazon, and many others, have tried different ways to control the information flow, creating a Web economy governed by search engines (Gerlitz and Helmond, 2013). Facebook newsfeed content has been optimized using EdgeRank, an interaction-based algorithm. This machine learning technique relies on several factors including affinity (e.g., how "friendly" a user is to a friend in his/her friend list, calculated using his/her time spent checking and interacting with that friend), type of content (e.g., relationship status update), and time (e.g., recent posts weigh more heavily than older ones) (Birkbak and Carlsen, 2016; Kincaid, 2010; Pariser, 2011).

Soon after online content accumulates, organizing it for users based on rankings is insufficient as there may be a gap between what users look for and how search results are ranked algorithmically. Many websites, therefore, allow users to manually manage the content they wish to be exposed to through various types of automatic recommender systems (e.g., alerts, Really Simple Syndication or RSS, and email sign-up, among others). Using these systems, users can select certain types of content from larger amounts of information according to their explicit preferences (Montes-García et al., 2013). Recommender systems have been heralded as one of the most powerful tools to help large online entities keep and grow traffic to their websites while putting those without similar techniques at disadvantageous positions in the competition (Hindman, 2018). Some examples of those that have been successful in providing their audiences a superior Web 2.0 individual user experience include YouTube, Google News, Yahoo News, and Facebook (Messing and Westwood, 2014). Not only do these user-customized systems work to avoid mismatch between what information users seek and what is returned by ranking algorithms, but the systems also help engage audiences, giving users some autonomy in personalizing the content they prefer.

But user-customized systems require some level of active participation from audience members, which does not universally apply to every user. As such, most commercial websites resort to tracking and learning about users' online behaviors in order to suggest what might be most desirable. Extant research found that 99 percent of the top 200 news sites have at least one type of tracker, and 50 percent of them include eleven types of user trackers (Kontaxis and Chew, 2015; Yu et al., 2016). Tech giants and news sites claim that tracking user behaviors is for the sole purpose of personalizing the content to users' "tastes" in information, despite accusations that user data are sold for commercial purposes (Lambrecht et al., 2014). A challenge for these tech giants then is to figure out how to make their content personally relevant to users. In order to do that, they follow users' online footprints ranging from search patterns, to purchase history, and to Web visits. Additionally, digital tech giants are offering various types of Web services as part of their strategies to expand their comprehensive control over the online population. Google, for example, rolled out Gmail, Google Groups, Google Drive, Google Maps, and Google Cloud, online services that require users to log in. It owns YouTube, sells Google Home, and often asks users to consolidate these services using their Google identity. In doing so, it has got its hands on enormous piles of user data (Pariser, 2011). Other Web entities are following suit. Facebook launched Messenger in 2011 (Bradshaw, 2011), acquired Instagram in 2012 (Hill, 2012), and bought WhatsApp for $19 billion (Covert, 2014). More recently, the company revealed its detailed plan of offering cryptocurrency, Libra, which, according to Facebook's claims, is a simple global currency aiming to serve billions of unbanked people worldwide (Constine, 2019; Godlewski, 2018; Libra, 2019). Comprehensive online packages like these further expand tech companies' digital services and also help gain a stronger understanding and control of users as richer and more user data become available.

For years, Facebook has been accused of manipulating what content its users can be exposed to in their newsfeeds. The tech giant has relied on algorithms to filter out information that is deemed irrelevant to individual users based on the data it gathers from their posts, interactions, and private messages (Marichal, 2016). The world's largest social network site has been criticized for its malicious utilization of user data for political marketing. For example, in 2016, Facebook sold data analytics of 50 million American individuals to the now notorious and defunct Cambridge Analytica, a firm that was linked to Trump's campaign adviser Steve Bannon. The data were then used to target Facebook users, feeding them with hyper-partisan information that allegedly escalated polarization among voters in the United States during the 2016 presidential election (Cadwalladr and Graham-Harrison, 2018). Facebook reportedly struck deals with other tech giants including Amazon and Netflix to further mine its user data for commercial marketing purposes without users' knowledge (LaForgia et al., 2019). To be effective in understanding what users want, multiple strategies have been deployed. They include users' location (Park et al., 2007), profile information (Chen et al., 2007), or click behavior (Liu et al., 2010). This dominant feature has influenced various sectors in society, from education to mobile application, to marketing, and many others (Dylko, 2016). Facebook's and other tech giants' practices of data collection and utilization have not been made transparent or explicit enough to many users. As such, from time to time, users are surprised about how many aspects of their lives are monitored and how much of their privacy has been turned into data points for commercial purposes by Facebook, Google, Amazon, and the likes (Fowler, 2019; Nichols, 2018; Roettgers, 2018).

Whether they are algorithmically driven or individuals' choice-based, these mechanisms are how tech companies respond to an important challenge in the digital age. That is: information abundance needs to be dealt with in ways that help make content manageable and consumable for Internet users. For users, the deployment of these mechanisms has given them an opportunity to control what information they want to be exposed to. For example, Liang and Fu (2017) found that Twitter users manage their information consumption on this social media platform by following certain groups of stable users. They will unfollow other users if they see the information they get from these users is redundant or overloading. For website owners, the goal of using algorithms to track users' online footprints is to increase traffic and user engagement, which is the currency of the business model.

It is important to note that how these systems work is mostly based on a simple psychological logic: People tend to prefer information that is congruent to their beliefs and values. Theoretically, these techniques help information consumption in the high-choice media environment by making it more efficient and interesting. There are, however, many pitfalls when the gatekeeping power shifts from editorial boards that use professional standards in selecting information to "opaque algorithms that are reined by their own logic, the logic of advertisers or consumers' personal preferences" (Möller et al., 2018, p. 2). As digital traces can be incomplete and exclusionary, gatekeeping, or more precisely curating information that is based mostly on such systems, has increasingly concerned scholars. It causes fear of negative impacts on the diversity and quality of information diffusion, leading to exacerbated polarization and creating an environment where misinformation thrives (Helberger et al., 2018). A Perspective Web would be welcome because it could make the mechanisms of the algorithms and the information they "push" more transparent to users. In the next sections, we will review several theoretical concepts (e.g., selective exposure, homophily, echo chamber, and filter bubble) that are related to the aforementioned problems.

7.3 Selective Exposure

Selective exposure has its origin in cognitive dissonance in the field of psychology. According to the cognitive dissonance hypothesis, the presence of dissonance, which causes psychological discomfort, will motivate people to avoid situations and information that would increase dissonance (Festinger, 1962). Based on this assumption, selective exposure theorists argue that in order to avoid incongruent information, people selectively choose to be exposed to messages that are congenial to their views while warding off incongruent opinions (Freedman and Sears, 1965). Another theoretical concept that corroborates this hypothesis is confirmation bias. After making a decision, we tend to try to find ways to confirm our decision-making by swaying away from information that proves otherwise while seeking information that supports our decision

(Festinger, 1962; Freedman and Sears, 1965). Research in the twentieth century found evidence supporting the selective exposure hypothesis on many issues such as car purchasing (e.g., Ehrlich et al., 1957), parenting techniques (e.g., Adams, 1961), personal care products (e.g., Mills, 1965), and political inclination (e.g., Schramm and Carter, 1959) (Stroud and Muddiman, 2012, for all of these citations see Stroud, 2008).

The past few decades have seen enormous changes in the areas of politics, digital technology, and mass media, creating an environment where research interests in selective exposure revive and thrive. At the beginning of the Internet, scholars warned that the prevalence of selective exposure practices among audience members might result in wider polarization and deeper partisanship across societies (Stroud, 2008; Sunstein, 2009). In the United States and Europe, for example, evidence showed increasing polarization (Prior, 2013), with partisan views being prevalent in many important public issues such as the war in Iraq, healthcare reform, gay rights, gun ownership, abortion, immigration, stem cell research, and global warming. Partisanship, having existed for many years in established democracies, has contributed to the divide of public opinion.

In terms of news media development, a comparison of the media environment in the past and today shows great changes, which are important factors in nourishing selective exposure. In the past, most news channels offered a mix of information that served the broader mass audiences rather than tailoring their content to individuals' viewpoints (Garrett and Resnick, 2011). According to Prior (2013), in its heyday, broadcast television was able to reach the less educated, providing them with news on politics and elections. In that limited media choice environment, no other alternative could take over the prime-time news people watch while having dinner. But cable TV and especially the Internet have created an escape for those who are less interested in news politics by turning toward entertainment programs.

In the Internet era, technological developments allow for more active news consumers, affording them the ability to exercise their control over what they want to read, listen to, or watch. Technologies also allow audiences to easily screen out the news they do not want to be exposed to, hence alleviating information overload. Scholars contend that the high media choice environment has changed the dynamics of news use by providing endless choices of sources to news consumers (Stroud and Muddiman, 2012). A growing number of online content providers has toughened the news market, fostering competition among news outlets in order to get audience attention, and when it is online: "eyeballs" (Tewksbury, 2005). The abundance of new sources has made it impossible to consume all the news in the vast sea of online information, or flip through hundreds of channels on TV. It therefore requires and also allows the audience to be selective in terms of choosing the content that they want to be exposed to. In short, "selective exposure, selective perception, and selective retention pervade the process by which we make sense of who we are as political creatures" (Jamieson and Cappella, 2008, p.75). Scholars, however, have warned of an adverse effect of "being selective" in information consumption: it reduces individuals' exposure to opinions that are different from their own (Dylko et al., 2017). This phenomenon contradicts the idea of a Perspective Web where users encounter diverse viewpoints.

7.4 Homophily, Echo Chamber, and Filter Bubble

Homophily refers to the pattern that comes into being when individuals seek to maintain or develop their social networks with similar people. Studies of homophily date back to ancient times when, in his best-known book on Nichomachean ethics, the philosopher Aristotle argued that people "love those who are like themselves." Subsequent scholars including Plato and modern sociologists noted the importance of similarities in how people grow their friendship networks. Lazarsfeld and Merton (1954) coined the term homophily to indicate the homogeneity of these networks. The two scholars also quoted the proverb "birds of a feather flock together," which has been used extensively in later research in the area of social network research (for more details, see McPherson et al., 2001). Algorithmic information filtering in the current digital media landscape has aided this tendency of interactions between different population groups by selecting the content produced or preferred by those with similar traits. Echo chamber, a concept that is directly related to homophily, describes situations in which one's voice is amplified and echoed back with similar sounds. Metaphorically, it represents a phenomenon in the information environment where people with the same interests and views seek to connect and interact primarily with one another in an enclosed space (e.g., online groups) (Sunstein, 2009). Members of these communities would claim and repeat information that conforms to their group's values or reinforces existing beliefs (Dubois and Blank, 2018). As such, "the more fully formed this network is, the more isolated from the introduction of outside views is the group, while the views of its members are able to circulate widely within it" (Bruns, 2017, p. 3).

The concept of the filter bubble was introduced by Eli Pariser, a political activist and tech entrepreneur, in his bestselling book *The filter bubble: How the new personalized Web is changing what we read and how we think.* To describe the term, Pariser (2011) used an anecdote about the gradual disappearance of posts, which is often ostensibly invisible, by his conservative friends in his Facebook newsfeed as the platform's algorithms figured he was politically left-leaning based on his clicking patterns. Such use of algorithms and the likes, according to Pariser, entrap users in filter bubbles, surrounding them with similar content. In other words, individual users fall into enclosed circles of information that are built either through their manual selection of the content they identify with or automatic algorithmic profiling done by websites.

Several aspects of these concepts need closer scrutiny. For example, are homophily, echo chambers, and filter bubbles limited to online environments only? How different are they conceptually? Are their effects on user information seeking real? First, in principle, all three concepts are not limited to only online environments but encompass offline interactions as well. Homophily and echo chambers, for example, underly the fact that people often find it hard to be surrounded by those with opinions that are different from their own, whether it is offline or online. The tendency of seeking similar others helps them avoid that. Modern platforms and the availability of recommendation systems have made enclosures and disconnections easier and more visible (Bruns, 2017). There is a key difference between the concepts of homophily and echo chambers on the one hand and filter bubbles on the other: while homophily and echo chamber are about individuals deliberately seeking to connect and build networks with like-minded others and avoid differences, filter bubbles emerge as the result of homophily and echo chamber (Bruns, 2019). According to Bruns (2019), clear definitions of these concepts are needed to move them "away from simple binary assessments (users either are caught in an echo chamber or filter bubble, or they are not) and towards the measurement of a user's degree of 'chamberness' or 'bubbleness' – that is, of their communicative enclosure" (p. 24). Some features of a few modern platforms allow for cross-posting beyond the boundaries of the echo chamber a user is part of. On Twitter, for instance, users can mention accounts they do not follow, thus carrying the views beyond their circle of peers, or retweet posts from those outsides of their networks and make them visible to other members of their echo chamber. Much discussion in this area among political pundits and scholarly communities has, so far, revolved around the consequences these phenomena may engender: the lack of diversity and quality in information diffusion, which would result in deleterious implications on democratic participation. For example, the European Commission (Caucaso, 2013, p. 27) raised its concerns about potential negative effects of filter bubbles to the public:

> Increasing filtering mechanisms make it more likely for people to only get news on subjects they are interested in, and with the perspective they identify with. There are

benefits in empowering individuals to choose what information they want to obtain, and by whom. But there are also risks. …the internet will enable people to be less engaged in society, given the increasing capabilities for personalized filtering and the decreasing presence of "general-interested intermediaries" (such as newspapers). Such developments undoubtedly have a potentially negative impact on democracy. Thus, we may come to read and hear what we want, and nothing but what we want.

An informed citizenry has been seen as an important component of any democracy, because, as (Pariser, 2011, p. 5) puts it: "Democracy requires citizens to see things from one another's point of view, but instead we are more and more enclosed in our own bubbles. Democracy requires a reliance on shared facts; instead, we are being offered parallel but separate universes." Other scholars have argued that isolating users in their information cocoons can turn moderate views into extreme ones, exacerbating polarization (Nguyen and Vu, 2019), which according to Zeynep Tufekci, as quoted by the *Economist* (2017): "It's like you start as a vegetarian and end up as a vegan." As being in an echo chamber or filter bubble can harden people's political views, another related concern about the dystopian visions of online segregation is the erosion of trust in the editorial authorities of the news media, which would allow for unverified information to reign in the new media ecology (Moeller and Helberger, 2018). The results of the two seismic political events of 2016, in which American voters elected Trump to be their president and the British voted to leave the European Union, have fueled extensive discussions on the negative effects of echo chambers and filter bubbles (Nguyen and Vu, 2019). In addition to privacy concerns, many have blamed social networking sites such as Facebook for allowing the spread of fake news, which fomented political extremism through feeding people with hyper-partisan misinformation (Guess et al., 2018). Obama, in his farewell speech in 2017, warned the American public of democracy being compromised as the result of us becoming "so secure in our bubbles that we accept only information, whether true or not, that fits our opinions, instead of basing our opinions on the evidence that's out there" (Landy, 2017).

More recent scholarly accounts have challenged the echo chamber and filter bubble thesis. Nguyen and Vu (2019, p. 2), for example, argued against the oversimplification of the echo chamber phenomenon, which tends to "subsume social news audiences to a very passive role – merely as 'lumps of clay' easily molded by algorithms." Bruns (2019) voiced similar concerns about the tendency of sidelining human agency in information seeking and consumption while assuming users' information diets are dictated by all-powerful algorithms. This, according to (Bruns, 2019, p. 19), "represents a dangerous slide toward a technologically determinist understanding of the contemporary media environment."

Empirical studies have found evidence that supports the more nuanced arguments. For example, Messing and Westwood (2014) found that social news users are more likely to read the news their friends share even if that news does not match their political ideologies. Flaxman et al. (2016) analyzed Web browsing data from 50,000 US-located users to find that the use of search engines and social media channels contributed to an increase in individuals' exposure to content from their less preferred side of the political spectrum. In his analysis of Twitter data from 225,000 accounts with more than 1,000 followers, Bruns (2017) found limited evidence of the formation of echo chambers in the Australian Twittersphere. Although the 225,000 Twitter accounts were scattered into different clusters, there were substantial interactions between them. According to the findings from their three-wave survey during the 2016 presidential election in the United States, Beam and colleagues (Beam et al., 2018) saw that Facebook use for news was associated with depolarization instead of polarization. Additionally, those who used Facebook for news were more likely to view both pro-and counter-attitudinal content. Using the results from a survey in Britain, Dubois and Blank (2018) found that individuals who were interested in politics and those who consumed news from a variety of sources had the ability to avoid echo chambers. As such, they argued, the fears of politically partisan segregation or the emergence of echo chambers might have been exaggerated.

Even though mixed empirical results have been documented, it would be erroneous to infer that the thesis about the existence of online homophily, echo chambers, and filter bubbles should be dismissed. Himelboim et al. (2013) examined Twitter clusters on ten public affairs issues and found discussions on these topics on these social networking platforms to be highly partisan, with users being unlikely to view cross-ideological content from the clusters of users they followed. Analyzing 150 million tweets from about 50 million users, Barberá and his colleagues (Barberá et al., 2015) found similar results, which indicated that information exchange happened primarily among users of similar ideologies. Conflicting findings have demonstrated that any evidence, either supportive or unsupportive of the notion of online segregation, is far from being conclusive, suggesting that a Perspective Web would be helpful, especially in giving people the opportunity to understand the perspectives behind information and its sources.

7.4.1 Do Digital Technologies Segregate Us?

Scholars, especially those who disagree with "the all-powerful algorithms" argument, have chased evidence for the important question: Are digital technologies the culprits that entrap people in filter bubbles or surround them with like-minded peers in echo chambers, aggravating polarization? In his book *Are filter bubbles real?* Bruns (2019) challenged the filter bubble and echo chamber thesis. He argued that building ourselves information cocoons with like-minded others would require extreme homophily with intense measures to control our exposure to different types of content. Some of them include cutting off communications with heterodox sources, "severing any existing contacts to non-adherents, online or offline," or closed and carefully controlled Facebook accounts to avoid groups and public pages, which "would be exceptionally time consuming and labor-intensive" (Bruns, 2019, p. 48). According to Bruns, if avoiding uncongenial information in today's media ecology was desired, users would need radical measures to succeed.

In the same line of thought, Dubois and Blank (2018) criticized studies on online echo chambers and filter bubbles for focusing on the use of one single digital platform to measure media diets. In this high-choice media environment, individuals may expose themselves to diverse information and perspectives either passively or actively. Dubois and Blank (2018, p. 740) noted:

> Whatever may be happening on any single social media platform, when we look at the entire media environment, there is little apparent echo chamber. People regularly encounter things that they disagree with. People check multiple sources. People try to confirm information using search. Possibly most important, people discover things that change their political opinions. Looking at the entire multi-media environment, we find little evidence of an echo chamber. This applies even to people who are not interested in politics. Thus, the possibility of being in an echo chamber seems overstated.

Robust evidence has lent support to this argument. For example, according to findings by Barberá et al. (2015), online social networks provided more opportunities for people to form and strengthen weak ties, which offline networks could not. Therefore, using social media platforms helped foster political diversity. Investigating the formation of online echo chambers among voters in Germany and Italy, Vaccari et al. (2016) found that exposure to congenial, oppositional, or mixed messages was associated with the structure of users' offline networks. The findings from these studies attest to the fact that the online network structure of connections through which users are exposed to information and perspectives mirrors their offline networks. This means even the evidence of polarization found online simply reflects what happened offline.

7.5 Political Polarization and Audience Fragmentation

Over the past decades, political polarization has been aggravated in various parts of the world. In the United States, where the political structure is built on a partisan system, records showed that differences in the public's views on several prominent issues, including government, race, immigration, national security, and environmental protection, among others, had sharply increased between 1994 and 2017 (Center, 2017). Specifically, in 1994 about 39 percent of Democrats and 26 percent of Republicans believed that "immigrants strengthen the country with their hard work and talents." In 2017, it rose to 64 percent of Democrats and declined to 14 percent of Republicans. In Europe, party polarization patterns have been observed in strong growth for the far right. For example, between 1980 and 2016, the number of European Parliament seats held by the far right increased from 2 percent to nearly 20 percent (Groskopf, 2016). In the latest election in 2019, the number rose to 25 percent. Countries that have seen greater polarization include Poland, Italy, France, Hungary, and Bulgaria (Erlanger and Specia, 2019). In Asia and Oceania, greater party polarization has been found in established democracies (e.g., Japan and New Zealand) (Dalton and Tanaka, 2007).

Audience fragmentation describes situations where people use a certain number of information sources and share with others in their like-minded groups. The high-choice media environment, where audience members are inundated with various sources of information, has triggered fragmentation as information overload and selective exposure are prevalent (Lee, 2007). Scholars have documented a growing trend of audience fragmentation (Prior, 2013). However, there is no supportive evidence of whether the online audience is more fragmented than offline information users (Fletcher and Nielsen, 2017), which rejects the notion that digital technologies are the culprit to divide audience members. Despite mixed scholarly findings of whether selective exposure, homophily, filter bubbles, or audience fragmentation exist in the modern media environment, providing users with sufficient details of any piece of information has always been encouraged. Doing so would be even more beneficial to audience members in the context of rampant falsehoods, which require users to have a certain set of skills to differentiate between facts and non-facts. A Perspective Web, where transparency is prioritized, would be helpful to consumers.

7.6 Concluding Remarks

The myth about algorithms causing online echo chambers or filter bubbles perhaps began with concerns about changes in the political and social landscapes that have not been adequately explained. Echo chambers, filter bubbles, and online information cocoons offer easy technological answers to problems that have recently emerged in many democratic nations (Bruns, 2019). Another important point to note here is that so far evidence of political polarization, which has been seen as the result of echo chambers and filter bubbles, has mostly been discovered in democratic societies. In these countries, expressing opinions or perspectives is not only protected by laws but also encouraged in multiple settings. In other words, having different opinions or perspectives, which drives polarization, is critical to any pluralist system. Encountering views that are oppositional to our own is almost inevitable both online and offline, out of or within our social networks. Bruns (2019, p. 50) argued that in the modern media ecology, selective exposure and homophily still exists as people need to organize their daily activities based on their existing network. However, "we cluster, but we do not segregate." Blaming the Internet and social media platforms is, therefore, easy but unhelpful for us to understand the complexity of recent political changes in various parts of the world.

That is not to say that online entities are not trying to manipulate online users using algorithmic-driven strategies because users' stickiness and advertising dollars will continue to be the incentive for their business models. It is, however, often difficult to know how their technical and social mechanisms work. For example, in its recent investigative article, *The Wall Street Journal* provided details on how Google is exerting its editorial control on user search result rankings in unclear ways by asking contractors to adjust algorithms. Such findings "undercut one of Google's core defenses against global regulators worried about how it wields its power" to interfere in users' content exposure (Grind et al., 2019, p. 15). To prevent any technical schemes that may lead to manipulating Web content or biased representation of different perspectives, scholars have suggested institutionalizing and incorporating transparency into how these mechanisms are operated (Helberger et al., 2018). Requiring websites to increase information diversity by making views from minority groups more prominent is another viable solution (Helberger, 2011). Holding these Web entities accountable for the spread of misinformation and conspiracies on their platforms would contribute to combatting extremism on the Internet (Bruns, 2019). For individual users, diversifying our media diets to meddle with tracking systems would help ensure that we are exposed to various types of content (Pariser, 2011). Although a Perspective Web is seemingly technologically possible, such an ideal digital environment will require changes in the business model of tech companies as well as strong ethical and social commitments to facilitate the free and factual exchange of information and perspectives.

References

Adams, J. S. 1961. Reduction of cognitive dissonance by seeking consonant information. *The Journal of Abnormal and Social Psychology*, **62**(1), 74.

Bakshy, E., Messing, S., and Adamic, L. A. 2015. Exposure to ideologically diverse news and opinion on Facebook. *Science*, **348**(6239), 1130–1132.

Barberá, P., Jost, J. T., Nagler, J., Tucker, J. A., and Bonneau, R. 2015. Tweeting from left to right: Is online political communication more than an echo chamber? *Psychological Science*, **26**(10), 1531–1542.

Bawden, D., and Robinson, L. 2009. The dark side of information: Overload, anxiety and other paradoxes and pathologies. *Journal of Information Science*, **35**(2), 180–191.

Beam, M. A., Hutchens, M. J., and Hmielowski, J. D. 2018. Facebook news and (de)polarization: Reinforcing spirals in the 2016 US election. *Information, Communication & Society*, **21**(7), 940–958.

Birkbak, A., and Carlsen, H. 2016. The world of edgerank: Rhetorical justifications of facebook's news feed algorithm. *Social Science Research Network , SSRN Scholarly Paper No. ID 2764210*, 2764.

Bradshaw, T. 2011. Facebook launches Messenger app. *Financial Times.*

Bruns, A. 2017. Echo chamber? What echo chamber? Reviewing the evidence. *Biennial Future of Journalism Conference*, FOJ17.

Bruns, A. 2019. *Are filter bubbles real?* Medford, MA: Polity Press.

Cadwalladr, C., and Graham-Harrison, E. 2018. Revealed: 50 million Facebook profiles harvested for Cambridge Analytica in major data breach. *The Guardian.*

Caucaso, O. B. 2013. *A free and pluralistic media to sustain European democracy.* Brussel, Belgium: European Commission.

Center, Pew Research. 2017. Political polarization, 1994–2017. *Pew Research Center*, November.

Chen, T., Han, W.-L., Wang, H.-D., Zhou, Y.-X., Xu, B., and Zang, B.-Y. 2007. Content recommendation system based on private dynamic user profile. *2007 International Conference on Machine Learning and Cybernetics*, **4**, 2112–2118.

Constine, J. 2019. *Facebook announces Libra cryptocurrency: All you need to know.* California: TechCrunch.

Covert, A. 2014. *Facebook buys messaging service WhatsApp for $19 billion.* Georgia: CNN.

Dalton, R. J. and Tanaka, A. 2007. The patterns of party polarization in East Asia. *Journal of East Asian Studies*, **7**(2), 203–223.

Dubois, E. and Blank, G. 2018. The echo chamber is overstated: The moderating effect of political interest and diverse media. *Information, Communication & Society*, **21**(5), 729–745.

Dylko, I. B. 2016. How technology encourages political selective exposure. *Communication Theory*, **26**(4), 389–409.

Dylko, I. B., Dolgov, I., Hoffman, W., Eckhart, N., Molina, M., and Aaziz, O. 2017. The dark side of technology: An experimental investigation of the influence of customizability technology on online political selective exposure. *Computers in Human Behavior*, **73**, 181–190.

Economist, 2017. Once considered a boon to democracy, social media have started to look like its nemesis. November.

Ehrlich, D., Guttman, I., Schönbach, P., and Mills, J. 1957. Postdecision exposure to relevant information. *The Journal of Abnormal and Social Psychology*, **54**(1), 98.

Erlanger, S., and Specia, M. 2019. European Parliament elections: 5 biggest takeaways. *The New York Times.*

Festinger, Leon. 1962. *A theory of cognitive dissonance.* Vol. 2. California: Stanford University Press.

Flaxman, S., Goel, S., and Rao, J. M. 2016. Filter bubbles, echo chambers, and online news consumption. *Public Opinion Quarterly*, **80**(S1), 298–320.

Fletcher, R., and Nielsen, R. K. 2017. Are news audiences increasingly fragmented? A cross-national comparative analysis of cross-platform news audience fragmentation and duplication. *Journal of Communication*, **67**(4), 476–498.

Fowler, G. A. 2019. Alexa has been eavesdropping on you this whole time. *Washington Post.*

Freedman, J. L., and Sears, D. O. 1965. Selective exposure. Pages 57–97 of: *Advances in Experimental Social Psychology*, vol. 2. Amsterdam, The Netherlands Elsevier.

Garrett, R. K., and Resnick, P. 2011. Resisting political fragmentation on the Internet. *Daedalus*, **140**(4), 108–120.

Gerlitz, C., and Helmond, A. 2013. The like economy: Social buttons and the data-intensive Web. *New Media & Society*, **15(8)**, 1348–1365.

Godlewski, N. 2018. Facebook owns a ton of popular apps, here are how a few big ones use your information. *Newsweek.*

Google. 2019. *About. California*, United States: Google.

Grind, K., Schechner, S., McMillan, R., and West, J. 2019. How Google interferes with its search algorithms and changes your results. *The Wall Street Journal*, **15**.

Groskopf, C. 2016. European politics is more polarized than ever, and these numbers prove it. *Quartz.*

Guess, A., Nyhan, B., and Reifler, J. 2018. Selective exposure to misinformation: Evidence from the consumption of fake news during the 2016 U.S. presidential campaign. Brussels, Belgium: European Council.

Heffernan, V. 2017. Just google it: A short history of a newfound verb. *WIRED*, **5**(November), 2019.

Helberger, N. 2011. Media diversity from the user's perspective: An introduction. *Journal of Information Policy*, **1**, 241–245.

Helberger, N., Karppinen, K., and D'Acunto, L. 2018. Exposure diversity as a design principle for recommender systems. *Information, Communication & Society*, **21**(2), 191–207.

Hill, K. 2012. 10 reasons why Facebook bought Instagram. *Forbes, Tech*, **12**(November), 2019.

Himelboim, I., McCreery, S., and Smith, M. 2013. Birds of a feather tweet together: Integrating network and content analyses to examine cross-ideology exposure on Twitter. *Journal of Computer-Mediated Communication*, **18**(2), 154–174.

Hindman, M. 2018. *The Internet trap: How the digital economy builds monopolies and undermines democracy.* Princeton, NJ: Princeton University Press.

Jamieson, K. H., and Cappella, J. N. 2008. *Echo chamber: Rush Limbaugh and the conservative media establishment.* Oxford, UK: Oxford University Press.

Kincaid, J. 2010. EdgeRank: The secret sauce that makes Facebook's news feed tick. *TechCrunch*, **5**(November), 2019.

Kontaxis, G. and Chew, M. 2015. Tracking protection in Firefox for privacy and performance. *Proceedings Of Web2.0 Security and Privacy (W2SP), IEEE Computer Society, Taichung, Taiwan*, 2.

LaForgia, M., Rosenberg, M., and Dance, G. J. X. 2019. Facebook's data deals are under criminal investigation. *The New York Times*, **12**(November), 2019.

Lambrecht, A., Goldfarb, A., Bonatti, A., Ghose, A., Goldstein, D. G., Lewis, R., Rao, A., Sahni, N., and Yao, S. 2014. How do firms make money selling digital goods online? *Marketing Letters*, **25**(3), 331–341.

Landy, H. 2017. The "great sorting": In his farewell address, Obama names the danger in our social media filters. New York: *Quartz*, November.

Lazarsfeld, P. F. and Merton, R. K. 1954. Friendship as a social process: A substantive and methodological analysis. Pages 8–66 of: Berger, M. (ed), *Freedom and Control in Modern Society*. New York, NY: Van Nostrand.

Lee, J. K. 2007. The effect of the Internet on homogeneity of the media agenda: A test of the fragmentation thesis. *Journalism & Mass Communication Quarterly*, **84**(4), 745–760.

Liang, H., and Fu, K. 2017. Information overload, similarity, and redundancy: Unsubscribing information sources on Twitter. *Journal of Computer-Mediated Communication*, **22**(1), 1–17.

Libra. 2019. *A new global payment system*. New York: Libra Group.

Liu, J., Dolan, P., and Pedersen, E. R. 2010. Personalized news recommendation based on click behavior. Pages 31–40 of: *Proceedings of the 15th International Conference on Intelligent User Interfaces*.

Marichal, J. 2016. *Facebook democracy: The architecture of disclosure and the threat to public life*. New York, NY: Routledge.

McPherson, M., Smith-Lovin, L., and Cook, J. M. 2001. Birds of a feather: Homophily in social networks. *Annual Review of Sociology*, **27**(1), 415–444.

Messing, S. and Westwood, S. J. 2014. Selective exposure in the age of social media: Endorsements of Trump partisan source affiliation when selecting news online. *Communication Research*, **41**, 1042–1063.

Mills, J. 1965. The effect of certainty on exposure to information prior to commitment. *Journal of Experimental Social Psychology*, **1**(4), 348–355.

Möller, J. and Helberger, N. 2018. Beyond the filter bubble: Concepts, myths, evidence and issues for future debates. *University of Amsterdam*, **12**, 2019.

Möller, J., Trilling, D., Helberger, N., and van Es, B. 2018. Do not blame it on the algorithm: An empirical assessment of multiple recommender systems and their impact on content diversity. *Information, Communication & Society*, **21**(7), 959–977.

Montes-García, A., Álvarez-Rodríguez, J. M., Labra-Gayo, J. E., and Martínez-Merino, M. 2013. Towards a journalist-based news recommendation system: The Wesomender approach. *Expert Systems with Applications*, **40**(17), 6735–6741.

Nguyen, A., and Vu, H. T. 2019. Testing popular news discourse on the "echo chamber" effect: Does political polarisation occur among those relying on social media as their primary politics news source? *First Monday*, **24**(5).

Nichols, S. 2018. Your phone is listening and it's not paranoia. *Vice Media*, **10**(November), 2019.

Papacharissi, Z. 2002. The virtual sphere: The Internet as a public sphere. *New Media & Society*, **4**(1), 9–27.

Pariser, E. 2011. *The filter bubble: How the new personalized Web is changing what we read and how we think.* New York, NY: Penguin.

Park, C. S. 2019. Does too much news on social media discourage news seeking? Mediating role of news efficacy between perceived news overload and news avoidance on social media. *Social Media + Society*, **5**(3), 2056305119872956.

Park, M. H., Hong, J. H., and Cho, S. B. 2007. Location-based recommendation system using Bayesian user's preference model in mobile devices. Pages 1130–1139 of: *International conference on ubiquitous intelligence and computing*. Springer.

Prior, M. 2013. Media and political polarization. *Annual Review of Political Science*, **16**, 101–127.

Roettgers, J. 2018. Facebook admits to scanning private messages, releases privacy policy updates. *Variety*, **12**(November), 2019.

Schramm, W., and Carter, R. F. 1959. Effectiveness of a political telethon. *Public Opinion Quarterly*, **23**(1), 121–127.

Stroud, N. J. 2008. Media use and political predispositions: Revisiting the concept of selective exposure. *Political Behavior*, **30**(3), 341–366.

Stroud, N. J., and Muddiman, A. 2012. Exposure to news and diverse views in the Internet age. *ISJLP*, **8**, 605.

Sunstein, C. R. 2009. *Republic.com 2.0*. 2 ed. Princeton, NJ: Princeton University Press.

Tewksbury, D. 2005. The seeds of audience fragmentation: Specialization in the use of online news sites. *Journal of Broadcasting & Electronic Media*, **49**(3), 332–348.

Vaccari, C., Valeriani, A., Barberá, P., Jost, J. T., Nagler, J., and Tucker, J. A. 2016. Of echo chambers and contrarian clubs: Exposure to political disagreement among German and Italian users of twitter. *Social Media+ Society*, **2**(20563), 05116664221.

York, C. 2013. Overloaded by the news: Effects of news exposure and enjoyment on reporting information overload. *Communication Research Reports*, **30**(4), 282–292.

Yu, Z., Macbeth, S., Modi, K., and Pujol, J. M. 2016. Tracking the trackers. Pages 121–132 of: *Proceedings of the 25th International Conference on World Wide Web*. New York: *Association for Computing Machinery*.

Zhang, S., Zhao, L., Lu, Y., and Yang, J. 2016. Do you get tired of socializing? An empirical explanation of discontinuous usage behaviour in social network services. *Information & Management*, **53**(7), 904–914.

PART III

Mediating Perspectives

Section Editors: Julia Noordegraaf and Thomas Poell

8

The Mediation of Online Information

Julia Noordegraaf and Thomas Poell

The starting point of this section of the book is the idea that information is produced in networks of actors and technologies that shape the perspectives embedded in this information. Being the product of people and evolving technologies, these perspectives are never universal, although certain perspectives become dominant in particular periods as social realities (Laing, 1990). The question is how this process of mediation takes shape in online environments. How are specific perspectives constituted, stabilized, and potentially challenged? To what extent can we as scholars gain insight in to this process? And how can we contribute to enhancing the "quality" of online communication?

In this section introduction, we conceptualize the key dimensions of the process of online mediation. The first three chapters each focus on a particular dimension: Chapter 9 on the source and its encodings; Chapter 10 on channel politics and technicity; and Chapter 11 on content, form, and reception. Each dimension is crucial for evaluating specific information and understanding how particular "perspectives" are embedded in online communication. The concluding Chapter 12 on "quality and perspectives" develops a human–computer interaction approach for tracing perspectives in online documents via an assessment of their quality.

These interventions should be seen against the backdrop of the current information crisis. During the 2016 US presidential elections, the Brexit referendum, and again in the 2020 pandemic, it became clear that the major platform companies – most prominently Facebook, Twitter, and Google – struggle with the moderation of the vast amounts of (dis-)information shared through their platforms (Frenkel et al., 2020; Gillespie, 2018; Bennett and Livingston, 2018). As platforms have become central to public communication, these struggles have a large impact on how and what information circulates in public space. In the mass media–dominated public sphere, control over the quality and relevance of information was primarily exercised through "gatekeeping": "selecting, writing, editing, positioning, scheduling, repeating and otherwise massaging information to become news" (Shoemaker et al., 2009). Whereas contemporary mass media continue to gatekeep the news, in the increasingly platform-dominated media landscape this gatekeeping process no longer determines whether information can enter public space and become "news." Content selection on platforms primarily takes place after publication, through an intricate interplay among user activities, platform curation, and moderation practices (Van Dijck et al., 2018). This has opened the door to conspiracy theorists, propagandists, and click-bait fabricators, who produce content that triggers platform users and algorithms (Benkler et al., 2018).

There have certainly been efforts to combat the information crisis. Over the past years, all major platform companies have become more proactive in moderating disinformation. The social media ban of the at-the-time sitting US President Donald Trump in early 2021 is symbolic of this development. Moreover, around the globe a wide variety of fact-checking initiatives and organizations have been established to support the moderation of online content (Graves and Cherubini, 2016; Graves and Mantzarlis, 2020). These efforts, however, remain work in progress. The 2020 infodemic demonstrated that despite concerted efforts by leading platform companies, disinformation continues to proliferate. Particularly challenging is also that platform curation and moderation largely remain a black box, as the workings of these systems are considered trade secrets. Consequently, it is often unclear how particular content and perspectives become prominently visible.

Given the growing and potentially damaging societal impact of online communication, we urgently need a more thorough understanding of how online content circulates, how particular perspectives are embedded in this content, and how the quality of this content can be verified. The chapters in this section contribute to this objective by exploring the key dimensions of online communication. Before introducing these dimensions, we will first discuss why and how we need a network communication model rather than a linear model to understand this process.

The classic model of communication, conceived by mathematician and electronic engineer Claude Shannon (1948), is based on a systemic conception of communication: a sender encodes a message that is transmitted via a certain channel to be decoded for the receiver, who then may provide feedback to the sender, thus closing the loop. This model was later revised by linguist Roman Jakobson (1960), who added the influence of the context in which the message was created. Although this model already acknowledged that noise could affect the flow of communication and distort the content or delivery of the message, it has been criticized for reducing the complexity of mediated communication.

In particular it has been taken to task for ignoring the active agency of the various actors – senders, receivers, social and technical infrastructures, knowledge and belief systems, power structures – who influence the creation, dissemination, and interpretation of messages (Chandler, 1994; Maras, 2000). Digital platforms as key agents in the communication process further increase this complexity, necessitating a revision of the model of communication.

Media studies have always been concerned with how processes of mediation shape perspectives in communication. Faced with the disruptive changes in online communication, media scholars have increasingly adopted Bruno Latour's actor network theory (ANT) approach, which traces how human and nonhuman actors become entangled in the production of knowledge and meaning (Latour et al., 2005). ANT yields a view on communication as a process of meaning-making involving multiple heterogeneous actors who connect with each other in particular assemblages. The analysis of such actor networks helps to pinpoint at which moment which actors shape the content, form, and reception of perspectives in the online information space.

Chapter 9 by Eric Hoyt addresses perspectives at the level of the sources of online information. It aims to identify the parameters that indicate how the identity and intentions of the source of a message are encoded in the messages as they appear online. Which parameters allow us to reconstruct the encoding process that took place when the sender produced and circulated the message via particular online media? How do we account for the "noise" (technical, semantic, etc.) involved in the communication of information, complicating the identification of the source and the decoding of its intentions? How do human and technological actors become entwined and how do these steer our interpretation of the message?

Hoyt zooms in on one particular "actor" in online-mediated communication: metadata, the short, often textual, descriptions that guide our interactions with online media sources in both visible and hidden ways. He shows how metadata is never fully neutral and infused with the goals, ideas, and motivations of their creators. Comparing the metadata creation practices of two online collections in the domain of cinema and media studies, the Media History Digital Library (a digitized book and magazine collection) and PodcastRE (an archive of born-digital podcasts), Hoyt in this chapter opens the black box of information encoding and decoding and shows how we can negotiate the inherent bias and subjectivity that determine what information we find and how we interpret it.

Jonas Anderson Schwarz in Chapter 10 continues the inquiry into the mediation of online communication by critically exploring Facebook as a particular techno-commercial architecture. He starts by observing that Facebook is not simply a communication environment but a platform that exerts normative power. Through the design of its interface and its algorithmic sorting mechanisms, it shapes how users perceive the world. How Facebook exerts this normative power is, according to Schwartz, first and foremost premised on its corporate objectives. As a company that derives the bulk of its income from advertising, the platform is designed to maximize user engagement to facilitate data collection and targeted advertising. Its role as a platform for the exchange of news and information is subordinated to this commercial objective.

Schwartz makes clear that this has major ethical implications. Although Facebook, pressured by public backlash and moral outrage, has started to more stringently moderate disinformation and extremist content, the company remains mostly indifferent to the types of judgments and arguments exchanged by users. In this regard, it radically differs from the media organizations that have historically controlled public communication. Largely disregarding conventional journalistic norms and editorial practices, Facebook is ultimately agnostic to processes of truth-making. Schwarz emphasizes that this is particularly problematic from a civic perspective, as not only external observers but also Facebook insiders have only partial knowledge of how meaning and truth are produced through the platform.

In Chapter 11, Christina Neumayer discusses how the techno-commercial architectures of platforms, as examined by Schwartz, shape the visibility or invisibility of particular societal perspectives in online communication. Building on classic communication theory, she makes a basic distinction between content, form, and reception. Translating this classic conceptual scheme to the online environment, she focuses on activists' struggles for public attention. The research on these struggles suggests that social media data have become a rich source on political protest. At the same time, a lot of data remain out of sight and, consequently, particular perspectives remain invisible.

In terms of content, Neumayer notices that while the study of mass media visibility is relatively straightforward, the analysis of online visibility is hugely complicated. Not only do platforms algorithmically curate and moderate the visibility of content but activists also use a variety of strategies to either make their content prominently visible or obfuscate their own traces. Turning to form, the chapter observes that activists, just as they have done in relation to mass media, package their political grievances in formats – personalized, short, and highly visual – that work well on platforms. In turn, this triggers the concern that the platformization of protest can lead to a hollowing out of political activism. When protest communication is adapted to the form of the tweet, post, comment, GIF, or video, the question is whether particular protest perspectives and issues can still be communicated. Finally, Neumayer considers the reception of protest communication through digital platforms. As platform users not only receive but also actively share, like, comment, and flag content, the reception process has a major impact on questions of visibility. This process also shapes perspectives, as users constantly add, change, redirect, or remove content when commenting and sharing items. Thus, in their struggle for platform visibility, activists inevitably give up control over online communication to both platforms and users.

This section concludes with a chapter on how to capture perspectives in online information, based on traditionally defined quality parameters. The analysis of perspectives in online information is inherently connected to an assessment of the quality of information, understood as best meeting the needs of users. Traditionally, access to information has been regulated by institutional actors which provide clear frameworks for assessing its quality, such as the editorial policies of news media. The platformization of public communication has complicated these quality assessments, making it harder to identify perspectives and evaluate their quality in online information.

In Chapter 12, Davide Ceolin, Julia Noordegraaf, and Lora Aroyo introduce Multidimensional Online Information Quality (MOIQ): a tool for quality assessment of online data. Based on quality indicators developed in Library and Information Science – a sector with a longstanding commitment to upholding values of trustworthiness and reliability of publicly accessible information – the MOIQ tool aims to support the gatekeeping role of media institutions in the age of platforms. Combining human and machine evaluations of online documents, the tool allows users to assess the quality of online documents, considering their accuracy, precision, completeness, relevance, neutrality, and trustworthiness. Like news fact-checking tools, it aims to counter the destabilizing effects of the "disinformation order" (Bennett and Livingston, 2018).

References

Benkler, Y., Faris, R., and Roberts, H. 2018. *Network propaganda: Manipulation, disinformation, and radicalization in American politics*. Oxford University Press.

Bennett, W. L., and Livingston, S. 2018. The disinformation order: Disruptive communication and the decline of democratic institutions. *European Journal of Communication*, **33**(2), 122–139.

Chandler, Daniel. 1994. The transmission model of communication. *University of Western Australia. Retrieved*, **6**, 2014.

Frenkel, S., Alba, D., and Zhong, R. 2020. Surge of virus misinformation stumps Facebook and Twitter. *The New York Times*, 8.

Gillespie, Tarleton. 2018. *Custodians of the Internet: Platforms, content moderation, and the hidden decisions that shape social media*. New Haven: Yale University Press.

Graves, L. and Cherubini, F. 2016. The rise of fact-checking sites in Europe. Reuters Institute for the Study of Journalism, University of Oxford.

Graves, L., and Mantzarlis, A. 2020. Amid political spin and online misinformation, fact checking adapts. *The Political Quarterly*, **91**(3), 585–591.

Jakobson, R. 1960. Linguistics and poetics. Pages 350–377 of: *Style in language*. MA: MIT Press.

Laing, R. D. 1990. *The politics of experience and the bird of paradise*. London: Penguin Books.

Latour, B., et al. 2005. *Reassembling the social: An introduction to actor-network-theory*. Oxford: Oxford University Press.

Maras, S. 2000. Beyond the transmission model: Shannon, Weaver, and the critique of sender/message/receiver. *Australian Journal of Communication*, **27**(3), 123–142.

Shannon, C. E. 1948. A mathematical theory of communication. *The Bell System Technical Journal*, **27**(3), 379–423.

Shoemaker, Pamela J, Vos, Tim P, and Reese, S. D. 2009. Journalists as gatekeepers. *The Handbook of Journalism Studies*, **73**, 73–87.

Van Dijck, J., Poell, T., and De Waal, M. 2018. *The platform society: Public values in a connective world*. Oxford: Oxford University Press.

9

The Source and Its Encoding: Reflections on Metadata in Digitized and Born-Digital Media Collections

Eric Hoyt

9.1 Introduction

Our perspectives on the Internet sources we encounter are profoundly shaped by metadata. Metadata – which is generally defined as textual information that describes some type of content or information – guides our encounters with digital objects in ways that are both visible and hidden. The title of a publication, for example, or the name of an author may enhance or diminish an article's credibility. These are examples of metadata that are visible to the user. Meanwhile, the metadata encoded within digital sources operate in many other ways that are invisible to most users. Categories are only searchable or sortable if the metadata is stored in ways that can accommodate such operations. In other words, our very ability to retrieve information on the Internet depends upon the metadata that describes that information. If the goal of the Perspective Web is to produce "a layer that supports people in knowing what is out there and how to value it" and "help them to measure consistency and judge the quality of information," then its success depends heavily upon accurate and reliable metadata.

However, metadata descriptions are never completely neutral. Metadata are embedded with goals, ideas, and motivations. Search Engine Optimization (SEO) has become a thriving pseudo-science, and the widespread attempts by content creators to implement SEO mean that metadata keywords may have more to do with self-promotion and making money than accurately describing the content. Yet, using information experts for assigning categories to metadata does not guarantee neutral objective results either, because such categories may express biases as well. Assigning categories – whether based on genre, subject, or identity (race, gender, sexuality) – is always an exercise of power, embedded in cultural assumptions and social norms.

This chapter seeks to promote greater awareness for researchers as they engage online with metadata. My focus is on metadata and digital collections within the disciplinary fields I know best – cinema and media studies. I will offer examples from two projects I have worked on that engage with distinct types of media, metadata, and digital assets: the Media History Digital Library (a digitized book and magazine collection) and PodcastRE (an archive of podcasts). In my discussion, I will highlight some of the major differences I have encountered between working with digitized media collections compared to born-digital media collections. My hope, however, is that the discussion and analysis will be valuable for researchers working in other domains beyond cinema and media studies. My overriding argument is that no metadata schema will ever be perfect and necessarily requires critical thinking and interpretation on the part of the researcher. We train students to be critical readers of the text. We need to think critically about the descriptions of the text as well. Toward this end, before visiting any specific projects, it is worth reflecting on metadata as forms of cultural encoding.

9.1.1 Metadata: Just the Facts?

By their definition, metadata are supposed to be purely factual and descriptive. In the US legal context, metadata are typically not protected by copyright because facts and ideas are not copyrightable. The Digital Public Library of America contends, "the vast majority of metadata … is not subject to copyright protection because it either expresses only objective facts (which are not original) or constitutes expression so limited by the number of ways the underlying ideas can be expressed that such expression has merged with those ideas" (DPLA, 2013). This legal status is important, allowing libraries to share catalog records at a huge scale and avoid duplicating the same work. We need metadata, and we need metadata to be openly accessible and reusable. Yet anyone who has spent time creating metadata records knows that subjectivity and bias are embedded within the process. As numerous information studies scholars have pointed out, the Library of Congress subject headings – widely used by university libraries – generally take the primacy of white men as a default. For example, "Robert Frost" is cataloged under "Poets, American" without reference to gender or race, whereas "Maya Angelou" is listed under subjects including "African American women authors" and "African American authors" (Hardesty, 2019). A recent documentary film, *Change the Subject: Libraries, Labels, and Activism* (2019), chronicles the struggle of undocumented college students to remove the Library of Congress subject heading "illegal aliens" (Broadley and Baron, n.d.). It is an important story about marginalized communities pushing back against schemas – both in metadata and society at large – that dehumanize and criminalize them. As we will see, podcasting is a medium that gives creators the agency to define their own identities and generate their own metadata, creating its own set of affordances and limitations.

The credits that provide attribution for films and broadcasting programs are a prime example of the necessary fictions and embedded cultural values encoded within media products. Transcribing the end crawl of a summer Hollywood blockbuster can provide a wealth of information, especially if the individual names (e.g., "Reed Morano") and categories (e.g., "cinematographer," "gaffer," "director") are stored in a relational database. But the credits themselves are products of contractual agreements and industry protocols, not objective appraisals of effort and contributions. For Hollywood studio movies – which frequently have several writers work on them before they enter production – the Writers Guild of America (WGA) has the ultimate authority to say which individuals get a "written by" credit, who gets "a story by" credit, and who gets left out entirely. The WGA's policy is to limit the "written by" credit to no more than two writers, with limited exceptions. And it is the WGA-approved credits that users encounter when they run searches within the Internet Movie Database. As legal historian Catherine L. Fisk chronicles in *Writing for Hire: Unions, Hollywood, and Madison Avenue*, the WGA struggled for decades to achieve recognition for the contributions of screenwriters (Fisk, 2016). Publicly pretending that fewer writers worked on a movie than actually did is a trade-off for upholding the importance of screenwriters as a whole. If an audience watching the credits sees one or two people listed as the writers, they see the authors. If they see nine or ten names, they see members of a committee and the screenwriters' contributions as a whole get minimized.

As all of these examples demonstrate, the encoding of metadata is itself encoded in cultural assumptions, norms, and expressions of power. By acknowledging these structures and assumptions upfront, we can make better choices as both the developers of Web-based frameworks and the users of such systems. In the remaining sections, I will share my own work engaging with metadata within the Media History Digital Library and PodcastRE, calling particular attention to the continuities and differences between working with digitized sources and born-digital sources.

9.2 Metadata in Digitized Magazine Collections

The project that demanded I become fluent in metadata standards is the Media History Digital Library (see Figure 9.1), which I lead at the University of

Figure 9.1 Front cover of the March 1, 1919 issue of *Moving Picture World*. Digitized by the Media History Digital Library, from the collection of the Museum of Modern Art Library.

Wisconsin-Madison. Founded by David Pierce in 2009 the MHDL is an open-access collection of over 2.5 million pages of digitized books and magazines related to the histories of cinema, broadcasting, and recorded sound. We have sought to use digital technology to build comprehensive collections of important periodicals that did not exist (especially not in keyword-searchable form) before we got started. For example, no library in the world possesses a full print run of *Moving Picture World* (1907–1927), which has long been one of the most important sources for historians of silent cinema. However, by borrowing and scanning volumes of *Moving Picture World* from the AFI, Museum of Modern Art Library, and private collectors, the MHDL has assembled a complete digital edition of this important source – enabling researchers around the world to read and search the publication across colorful scans, rather than turn page-by-page through the poor-quality microfilm edition. The encoding of metadata is key to providing researchers with a sense of provenance as they engage with the materials online. Without metadata, we could not attribute which volumes of *Moving Picture World* came from the Museum of Modern Art Library and which ones came from private collectors. Digitization strips objects of their contexts; metadata can be a vehicle for restoring lost contexts and creating new contextual frameworks.

Let us continue with the *Moving Picture World* example as a way of thinking about how a 100-year-old source becomes digitally accessible. How much description, and what types of metadata, does this journal need for users to effectively engage with it online as a source? In the early years of the MHDL, we focused on scale over granularity, entering as little metadata as possible. We scanned massive bound volumes of *Moving Picture World* (frequently exceeding 1,000 pages), entering just fifteen lines of metadata to describe the entire volume. In other words, the three-month bound volume of "March–April 1919" would be cataloged as an individual object rather than an issue-level, page-level, or article-level approach to metadata entry. Our strategy was to use optical character recognition (OCR) to make the journals searchable, reckoning that our users would accept locating the bibliographic information necessary for a footnote as a trade-off for free and fast keyword searchability. Over the last two years, we have changed our approach, scanning and cataloging magazines on the issue-level rather than entering them as large bound volumes. The result is that users will increasingly see that their search query yielded a keyword hit in the "March 1, 1919" issue of *Moving Picture World* rather than "March–April 1919." This is clearly an improvement. But is it sufficient?

If we take the intervention of the Perspective Web seriously, then the MHDL needs to contribute more metadata to aid with the interpretation, and not just the description, of *Moving Picture World* and other digitized film journals. One step that we have already taken is to add a paragraph discussing the history of *Moving Picture World* that accompanies every keyword search hit. But a static paragraph of text cannot trace the dynamic evolution of a trade paper spanning two decades of transformative change within the film industry. There is a tremendous amount of accompanying information we could add alongside keyword matches: the circulation figures, readership trends, and biggest advertisers of *Moving Picture World* and other trade papers; box office grosses, the biggest stars, and major world events to better contextualize the industry that the papers were reporting on. Archivists have sometimes referred to these types of associated information as "contextual metadata" and "significant properties" (Marchionini et al., 2009; Stepanyan et al., 2012). Contextual metadata is essential if we want to move toward the goal of a Perspective Web.

Adding all of that contextual metadata, however, takes a great deal of time and work. And if you push the entry of contextual metadata to its limit, you may wind up writing a work of history. I have essentially reached this point in my own work. Rather than put all of my time and resources into entering contextual metadata within the Media History Digital Library, I decided to write a book about early Hollywood trade papers that tells the interlocking stories of *Moving Picture World* and its competitors. My book was published this year with an open-access license, making it freely downloadable alongside the sources it chronicles and analyzes. From a programming standpoint, this arrangement is low-tech. But I hope it advances the goals of the Perspective Web while reminding us, as E. H. Carr eloquently argued sixty years ago, that history emerges from the interpretation of sources, from a conversation between past and present, and not simply from the organization and presentation of facts (Carr, 2018).

9.3 Podcasts, RSS Feeds, and the Encoding of Born-Digital Objects

Whereas the Media History Digital Library produces metadata for digitized sources, PodcastRE (short for Podcasting Research) collects and presents the metadata of born-digital objects. Founded by my colleague Jeremy Wade Morris at the University of Wisconsin-Madison, PodcastRE seeks to preserve contemporary sound culture by creating a searchable and researchable archive of podcasts. Currently, the PodcastRE website – accessible at http://podcastre.org – contains over 2 million archived podcasts, along with tools for search and analysis (Morris et al., 2019). In my work developing the data analytics features of PodcastRE, I have been struck over and over again by the significant differences between digitized works and born-digital objects, especially when it comes to questions of agency and authority.

The technological backbone of PodcastRE – and, arguably, podcasting as a medium – is the RSS feed (Hansen, 2021). An XML-based protocol, RSS allows for podcasters to easily publish their completed work and distribute it to audiences, who can opt to subscribe to particular feeds. In many ways, the metadata and the open feed are what separate a podcast from other media files on the Internet, including other forms of on-demand audio (for example, music streaming platforms and audiobook companies). The RSS feeds contain the metadata that podcasters themselves choose to enter. Consequently, podcast metadata is messy, idiosyncratic, and incomplete – worlds apart from the familiar and relatively consistent metadata standards of TEI and Dublin Core. As Susan Noh has argued, podcasters can use this open-ended approach to metadata as an opportunity for self-definition, agency, and the creation of bottom-up networks (Noh, 2020). For example, when the creators of the PHX podcast entered the keywords "podsincolor" and "women of color" within their RSS feed, they actively chose to present themselves this way and place their work within a larger network of podcasts produced by people of color. Any application of the Perspective Web to the domain of podcasting needs to honor the medium's affordances for creator self-definition through metadata.

At the same time, however, podcasters frequently stuff their RSS feeds with keywords to enhance the discoverability and credibility of their work. Like other Web content creators, podcasters may try to the game the system of SEO to make their work appear more relevant in search queries. Perhaps you have heard the old Internet joke? An SEO specialist walks into a bar, pub, night club, watering hole, coffee shop, restaurant, café, house, office, nation, world. Ironically, due to changes in search algorithms, most of these SEO keyword-stuffing attempts no longer hold much weight when it comes to improving discoverability within commercial search platforms. But these SEO-oriented keywords – along with the abovementioned uses of keywords for creator agency and self-definition – become visible and reactivated in one of the data visualizations developed by the PodcastRE team.

PodcastRE's Associated Keyword Word Cloud allows researchers to search for a particular keyword (see Figure 9.2). The application then generates a data visualization of the keywords most frequently associated alongside it within PodcastRE's index of RSS feeds. To enhance the visualization's legibility, the word cloud's keywords are sized proportionally to the number of times they co-occur with the queried term. And to make the visualization more useful to researchers, the visualization is dynamic and interactive. Clicking on a keyword within the word cloud will generate a list of results of all the podcast episodes in which that keyword and the queried keyword co-occur. Figure 9.2, for example, provides a visualization of the keywords that frequently co-occur in podcast RSS feeds alongside the keyword "truth." Beyond the self-evident and most

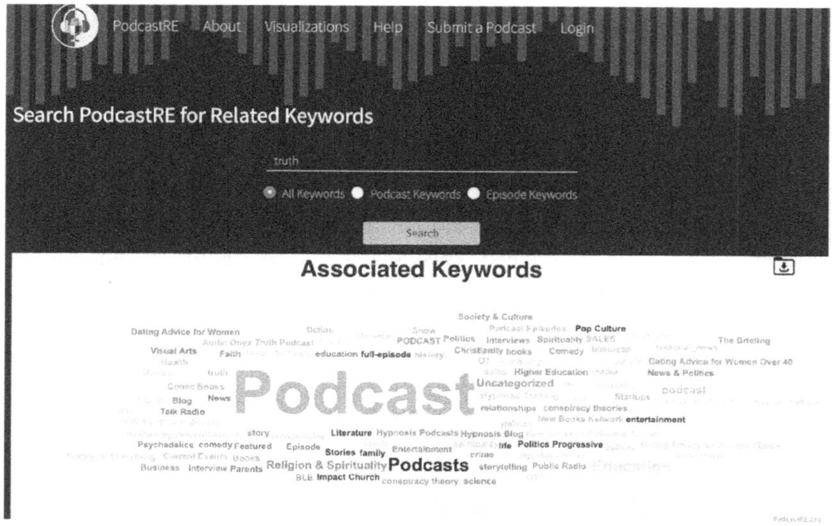

Figure 9.2 Associated Keyword World Cloud visualization for keyword, "truth," generated using the PodcastRE Analytics platform at https://podcastre.org/analytics and hosted by the University of Wisconsin-Madison.

prominent associated keyword ("Podcast"), the other keywords that co-occur with "truth" with some regularity include "education," "health," "spirituality," "conspiracy theory," and several variants of "dating advice for women."

With such a wide range of keywords encoded into RSS feeds, what can PodcastRE's Associated Keyword Word Cloud tell users about the claims of "truth" that we encounter in the podcasts and other born-digital sources? Some of the limitations are immediately obvious. The Associated Keyword visualization, in its current form, does not tell users that one health-themed podcast's claim to truth is greater or more accurate than another. In that sense, the visualization falls short of some of the goals of the Perspective Web, articulated earlier in this book in regards to health information and the vaccination debate (which dozens of podcasts address, some perpetuating harmful myths). Yet I think the visualization still holds a great deal of relevance for the Perspective Web. The takeaway of Figure 9.2 should not be that all health podcasts are equally valid in their information. Rather, the takeaway should be that "truth," "education," and "health" are keywords that podcasters mobilize in an effort to make their work more discoverable and seem more legitimate. By experimenting and clicking on different words within the word cloud, PodcastRE's users can begin to see the connections across podcasts and the networks in which podcasters (and their metadata) are embedded. These visualizations and interpretive exercises can help develop the critical thinking skills that users need for evaluating the sources and metadata encoding schemes that they encounter online.

9.4 Conclusion

As this chapter has shown, the sources we encounter on the Web are encoded with complex metadata. And, if the objectives of the Perspective Web are to become a reality, it will require the creation of more – much more – metadata. (What is a layer of information that guides people in evaluating the sources they find online if not new clusters of metadata?) The metadata model applied by libraries to digitized sources has the benefit of consistency and controlled vocabularies. Yet, as we have seen, they are also embedded with metadata bias, and their descriptions inevitably fall short in capturing how a dynamic textual object (like a magazine) changes over its lifespan. Born-digital objects, meanwhile, are typically encoded with the metadata entered by their creators. This opens up the possibilities for creator agency and self-definition that resist authoritative structures that assume the whiteness and maleness of creators. But it also opens the door to the stuffing of self-promotional keywords and questionable truth claims into such objects.

Where do we go from here? Clearly, the creation of better and more granular metadata remains a valuable goal. It must proceed, however, with a greater awareness of positionality. For those of us entering metadata for digitized objects, we need to ask ourselves: what social identity positions do we occupy, and how can we make sure we avoid reinforcing harmful structures of privilege and exclusion? We need to either update or do away with outdated schemas that assume the primacy of white men as their default.

Additionally, we need to promote greater metadata literacy among users. Critical thinking strategies are essential for interacting with metadata – sometimes visible, sometimes not – in digitized collections, movie credits databases, archives of born-digital objects, and other corners of the Web. The goals of the Perspective Web can only be achieved through greater awareness among both developers and users, metadata creators, and metadata consumers. The potential positive contribution of any new database, algorithm, fact-checker, or expert will be muted unless we promote foundational critical thinking and media literacy for Internet users.

Acknowledgments: Thank you to the American Council of Learned Societies, National Endowment for the Humanities Office of Digital Humanities, and the University of Wisconsin-Madison for supporting the projects described in this chapter. Thank you also to the many people who contributed to the projects, especially Jeremy Wade Morris, Samuel Hansen, Peter Sengstock,

David Pierce, Kelley Conway, Carl Hagenmaier, Wendy Hagenmaier, Andy Myers, Stephanie Sapienza, Derek Long, Tony Tran, Kit Hughes, Charles Acland, Kevin Ponto, Alex Peer, JJ Bersch, Susan Noh, Jacob Mertens, Matt St. John, Olivia Riley, Lesley Stevenson, and Connor Perkins.

References

Broadley, S., and Baron, J., directors. 2019. *Change the subject*. Dartmouth Digital Library Program. https://n2t.net/ark:/83024/d4hq3s42r

DPLA. 2013. *Policy statement on metadata*. Boston: DPLA Pro.

Fisk, C. 2016. *Writing for hire: Unions, Hollywood, and Madison Avenue*. Cambridge, MA: Harvard University Press.

Hansen, S. 2021. The feed is the thing: How RSS defined PodcastRE and why podcasts may need to move on. Pages 195–207 of: Morris, J. W. and Hoyt, E. (eds), *Saving new sounds: Podcast preservation and historiography*. Ann Arbor: University of Michigan Press.

Hardesty, J. 2019. *Bias and inclusivity in metadata: Awareness and approaches*. Indiana University Digital Collection Services.

Marchionini, G., Tibbo, H., Lee, C. A., Jones, P., Capra, R., Geisler, G., and Russell, T. 2009. *VidArch preserving video objects and context final report*.

Morris, J. W., Hansen, S., and Hoyt, E. 2019. The PodcastRE Project: Curating and preserving podcasts (and their data). *Journal of Radio & Audio Media*, 26(1), 8–20.

Stepanyan, K., Gkotsis, G., Kalb, H., Kim, Y., Cristea, A. I., Joy, M., Trier, M., and Ross, S. 2012. Blogs as objects of preservation: Advancing the discussion on significant properties. Pages 218–224 of: Moore, R., Ashley, K., and Ross, S. (eds), *Proceedings of the 9th International Conference on the Preservation of Digital Objects*. Toronto: University of Toronto Press.

10

Knowledge-Making on Techno-Commercial Platforms: The Example of Facebook

Jonas Anderson Schwarz

10.1 Introduction

Epistemology is the scholarly term for how humans consider the nature of knowledge. In my case, I advocate a realist view where empirical observations about the world will have to be made to make knowledge statements about this world. These observations are always perspective-bound and constrained by the observational tools at hand. In media studies in particular, epistemologies are generally saturated with folk aspects; practically everyone has a stake in defining the various ways in which the media contribute to knowledge-making in society, and practically everyone has accumulated various forms of knowledge through the media, about the media. This is important, particularly to media ethics (Ess, 2011; Srnicek, 2011), since communication technologies and organized media infrastructures are vehicles for "secondary socialization" and social norms for just about everyone. Not only scholars try to make sense of the media-saturated social world, but also ordinary citizens do, every day. Hence, the study of media epistemology has to consider various aspects of claim-making, including highly vernacular ones.

This chapter explores how theories of civic justification (Boltanski and Thévenot, 2006) and rationalities of ethical belief (Cheney et al., 2011) inform how citizens relate to Facebook as an entity in their everyday lives. My main argument is that individuals' orientation to Facebook is premised on those front-end interfaces through which Facebook becomes comprehensible to the user, and due to the opacity of these interfaces, user knowledge about Facebook is essentially abductive. In previous writings, I have expanded on some of the ways in which the digital realm relies on an epistemology of abduction (Andersson Schwarz, 2016), as users of digital infrastructures are always forced to "make do" with what the platform interface offers. The theory of abduction was initially outlined by semiotician Charles Sanders Peirce, referring to a particular kind of non-deductive inference that involves the generation and evaluation of explanatory hypotheses. In layman's terms, abduction is the capacity to make qualified guesses out of limited data. Luciana Parisi (2017) has broached the problem of abduction in relation to digital milieus, in a poignant footnote which she attributes to Whitehead (1929):

> Material computation, mainly relying on the inductive logic of physical interconnections, problematically omits the abstractions carried out by conceptual prehension for which there can be no direct observation, intuition or immediate experience. (Parisi, 2017, p. 97)[1]

In short, computers are brute, extremely reductionistic data processors, but very fast and very precise. To make sense of the data, humans are always required. In the case of Facebook, the ability of users to make abductive knowledge and ethical judgment is one step removed from the corporation's data analysis facility. Design decisions are made on behalf of users by service managers and designers, who use the data harvested to make "choice environments" (Thaler and Sunstein, 2008) which form the very core of the services on offer. The motivated reasoning of Facebook users can therefore be said to be shaped, or primed, by the algorithmic designs made for them.

Now, there is nothing inherently new to the claim that individual agents only have partial or limited accessibility to observe the social world. This goes for virtually all social environments. However, the example of Facebook, in particular, shows that this feature of social life becomes very pronounced and arguably very urgent when the social world in question is directly mediated through text and numbers. Measures like "likes" and plaintext postings give rather little scope for intuition or multisensory sensemaking; they are premised on objectivizing logics to begin with. They appear as artifacts, tokens, palimpsests of recorded social action – and it is in this capacity they risk reifying social observations, rendering actions as if they were objective statements or facts, thereby obscuring the always conditional, contextual, partial nature of such statements. Later, we will see that this normative power has both ethical and political dimensions.

10.2 The Normative Power of Prescribed Social Action

Media scholars tend to express concern for what I would label *conventional journalistic epistemological challenges*, where sites like Facebook constitute

[1] The concept of abduction was embraced also by pioneer cyberneticist McCulloch (1965), and later embraced by other cyberneticists like Heinz von Foerster, who represents a very empirical understanding of knowledge, where cognition is largely identified with all the processes that establish "meaning" from experience (von Foerster, 2003, p. 105). Magnani (2009) presents an up-to-date introduction to the concept of abduction.

"new venue[s] for a kind of publishing that is not mediated by any sort of vetting process" (Daniels, 2014, p. 143). More on these challenges later.

In addition to this conventional journalistic framing of epistemology, there are, however, also numerous other sources of knowledge about the world that are communicated through civic techno-commercial architectures like Facebook. Take, for example, normative social knowledge: Clues as to how one should behave as a person, what type of language one should use; images of what constitutes a crowd, or a critical mass of people; imaginaries of what at all *counts* as an argument, a fact, or a value. This may sound vague, but it is actually at the core of what it means to be a social human being; it is just that most of us make these decisions almost entirely unwittingly or automatically since knowledge like this becomes very integrated with the shared social fabric of everyday life.

Historians like Nikolas Rose and Ian Hacking have drawn on the work of Michel Foucault so as to explore how the very notion of what counts as a person, that is, the concept of *personhood*, has developed and shifted throughout history (Sugarman, 2014). For the purposes of this chapter, I will focus on Hacking in particular. He relates Foucault to analytical philosophy, offering a rather formalistic theory of being,[2] where Hacking, through his concept of "making up people" (Hacking, 2004, 99–114), refers to the ways in which category-making literally allows for certain ways of being to emerge. Note here the ontological dimension in data governance, where certain labels, categorizations, or, for that matter, the mere act of naming certain phenomena linguistically compel these things into being. He emphasizes, in particular, how practices of naming are interactive with what the social agents in question choose to name. Language can therefore be generative and transformative, informing and structuring the world, in a kind of "looping effect" (Hacking, 1995): When individuals change the ways in which they describe themselves, they change into different persons than they were before.

This is precisely why it is of such grave concern to explore the categorizations, classifications, and nominalizations that techno-commercial infrastructures like Facebook give rise to. The "looping effect" becomes particularly apparent when considering the entanglement of economics with cybernetics that lies at root of techno-commercial infrastructures of this kind. When aggregated, human behavior becomes, at least to a certain extent, predictable. As, for example, Thaler and Sunstein (2008) have pointed out, this is observably true in many ways. But perhaps it partly becomes true because people indeed

2 "My historical ontology is concerned with objects or their effects which do not exist in any recognizable form until they are objects of scientific study" (Hacking, 2004, p. 11).

are encouraged to act in certain ways on these platforms if we are to take the theories of prescription at face value.

It is an observed fact in cognitive psychology that when alternative readings of situations become salient, people tend to assign disproportional probability to the possible scenarios resulting from such salient alternatives. Gerken (2017) attributes this to something he calls *focal bias*. When humans are attuned to a particular eventuality, they automatically adapt their observations to it, as if it would be relevant to their knowledge-making – even when it is not. So, as soon as someone is labeled something, that will have a bearing on how people would treat the utterances of that person, regardless of the actual qualities of this person or of the utterance in question. Similarly, it is not hard to think that the same bias would arise in respect to artifacts and infrastructural properties. If a particular signal from Facebook is presented as being representative of a larger population, observers will assign weight to it, based on the assumption that it ought to be representative and thus important – regardless of whether it actually is representative. The question is once again actualized: whether the individual user can know how the information presented to her is in any way representative at all and, in that case, how (Schou and Farkas, 2016, p. 42).

Moreover, if Hacking's "looping effect" is true, this would also have a bearing on people's understanding of their own selves. If certain features of human beings are highlighted, made apparent, and perhaps even appear to be measurable – say, one's number of followers online, and so on – it is reasonable to expect that this would have a bearing on how important individuals estimate their own utterances to be, for example. Especially so, given that the numbers of "likes" and so on are often the only real cues as to how well a particular utterance was received since there is no real overseeability or access to more detailed metrics to begin with. This prompts the question of "the underlying, algorithmic mechanisms governing what is [at all] deemed information" (Schou and Farkas, 2016, p. 42) to begin with, and what is not; the "knowledge logic" at hand (Gillespie, 2014, p. 168). How and why is it presented to the user? Interestingly, the same epistemological tangle also faces those professional users of Facebook, operating the functional backend of the infrastructure, namely those who interpret user data and are in the position to make out of these data.

As scholars of the darker aspects of social media campaigning have demonstrated (Benkler et al., 2018; Farkas and Schou, 2019),[3] the popularity bias (Webster, 2014) of social media platforms means that collective conformism

[3] Key terms here are "fake news," disinformation, gaslighting, cyberbullying, and hate speech – but also less obvious, more quietly normative aspects like "virtue signaling" and different forms of moralism, including so-called "call-outs" where attention is directed to particular behaviors that are identified as lamentable.

is never very far away. Indeed, the infrastructures in question are designated to steer users in various ways and prescribe certain behaviors. Academic concepts and theories like *affordances* (Gibson, 1979), *grammars of action* (Agre, 1994), *code-as-law* (Lessig, 1999), *technical prescription* (Slack and Wise, 2002), and *nudging/libertarian paternalism* (Thaler and Sunstein, 2008) all attest to this, and it is important to note that it is not only the physical setup of interfaces and tools that shape social action but also attendant narratives and semiotic interventions made to help shape *sociocultural imaginaries* about the infrastructure in question (Jasanoff, 2015; Mansell, 2012). Facebook has normative power not only as environment but as a story about the world and what qualities this world is premised upon. Thankfully, such normative power is never absolute. It is encoded by producers, but interpretations might differ from user to user, or between certain groups or classes of users, for that matter.[4]

10.3 The Corporate Mission

This brings us to the corporate narratives of Facebook. Facebook's primary corporate goal is to expand shareholder value. Since its bulk of global revenue (98.5 percent) comes from advertising, the corporation's means of expansion are highly premised on maximizing user engagement with the corporation's properties so as to let its users become exposed to and engage with as many ads as possible and/or maximizing the revenue from each ad interaction. Arguably, any other goal of Facebook as a corporation – such as, for example, generating knowledge (mapping user interests so as to be able to predict user habits and optimize the targeting of its advertising) – becomes secondary to this overarching capitalist goal. In public consultation, the corporation's representatives normally reject this narrative, preferring instead to highlight other features of the company's ethos.[5] This discrepancy within the company's own narratives about the platform is particularly interesting.

The phrase *"Making the world more open and connected"* comes to mind – a mantra that Facebook executives used to repeat for as long as this remained the corporate "mission statement" of Facebook's head office, so as to steer the interpretation of what Facebook is, and why it exists.[6] But on June 22, 2017, Facebook CEO Mark Zuckerberg revealed a new mission statement, to *"Give*

[4] This is concurrent with Stuart Hall's (Hall, 1980) much earlier theory of normative mass media bias.

[5] E.g., in my own public consultation with Facebook's Public Policy Manager for the Nordic region, Christine Grahn, November 11, 2020.

[6] I have personally heard this in meetings with Facebook executives, on numerous occasions.

people the power to build community and bring the world closer together." Why this came to be is not entirely clear (this could be the subject of a full article, a book even), but it points to the inherent instability[7] of what a thing like Facebook is thought to be. Both of these accounts are similar, however, in that they present Facebook's primary goal as one of social connection. I therefore choose to label this a connectionist corporate ethos.

What observers might think of as "bare-bone," purely techno-commercial infrastructures are in fact almost never presented as such to the public: interpretations of infrastructures are almost always closely intertwined with semiotic statements "steering" the initial prescriptive force of the technology in different directions, much like the captioning text steers the interpretation of a photographic image (Barthes, 1964). Lisa Parks has argued that infrastructure has *intelligibility* as part of a general "phenomenology of infrastructure" (Parks, 2012, p. 67). That is why mission statements like these are important; they help to steer the publicly shared, collective perspectives of the technical systems in question. If one, for example, presents an entirely different narrative of Facebook as a technical system, as I set out to do in my biosemiotic reading of Facebook (Andersson Schwarz, 2018), the same "bare-bone" infrastructure comes to appear in a different light.

What is more, when digital infrastructures become so widely adopted that they become de facto industry standards (see Plantin et al., 2018, for an argument that digital platform businesses are in many ways infrastructure providers), this type of prescription also comes to have organizational aspects, as, for example, mass media producers adapt their entire management principles to being compatible with the infrastructure in question. In the case of Facebook, this has been pointed out by Caplan and Boyd (2018) and Helmond (2015). They show how Facebook's normative connectionism is imbricated in the understandings that corporate executives all over the world tend to use as arguments for having to connect their organization with the Facebook infrastructure and align it as such. Scholars of public-service media management have addressed this too (Andersson Schwarz, 2016) since it is a dilemma for public-service actors in particular: Should such media systems willingly give assent to Facebook and YouTube by, for example, linking their material to these companies' properties, or by posting public-service-financed media content on these platforms, or by integrating their services with user authentication/identification services provided by Facebook or Google?

[7] In Andersson Schwarz (2018) I use the term "metastability," borrowed from Gilbert Simondon.

Today's digital platforms are mediators that "exhaustively analyze, repackage, and then republish customer information [so as to] elicit as much information as possible" (Sylvain, 2018, p. 274). Through "algorithmic interpolation" (Cheney-Lippold, 2017, p. 113), the corporation makes *probable categorizations* that come to inform the design and maintenance of Facebook as an infrastructure. Due to its popularity, it so happens that many of Facebook's properties (remember, also Instagram and WhatsApp are owned by the company) are nowadays largely co-constitutive of human, everyday subject-making, since these media infrastructures saturate so many people's everyday life. It has thus been argued, from a regulatory point of view, that the architectures in question – platforms that enable human exchanges yet mediate, indeed intervene in the nature of the exchange taking place – should be judged in relation to the extent to which they actually intervene in their users' action.[8]

In terms of perspectives, this is important since the algorithmic infrastructure at the technological core of Facebook (mathematic operations of humanly unimaginable machinic speed and volume) "control the 'visibility' of friends, news, items, or ideas" (van Dijck and Poell, 2013, p. 49) with the intent of "making the content more relevant to its potential consumers" (O'Callaghan et al., 2015, p. 460). Here, the ad-based financing model once again becomes pertinent: Unlike platforms like Wikipedia, with entirely different business models and service designs, the primary goal of an infrastructure like Facebook is not to generate knowledge or, to any reasonable extent, "reliable" information. The explicit purpose of Facebook is to maximize user engagement so as to expose users to as much advertising as possible without harming the retention of user attention (Wu, 2016). The Facebook designers expend considerable resources and efforts in the service of the balancing act of making the user experience as pleasurable and tempting as possible while at the same time maintaining the profit-generating purpose of exposing users to ads. Consequently, the corporation's connectionist ethos would be ancillary to this ultimate goal. The implications for perspective-bound knowledge-making on and through Facebook should be obvious: The cards are not stacked in favor of objectivity, in any meaningful scientific way. Rather, the infrastructure is built for maximizing the rate of interaction between it and its users and, hence, privileging social circulation more than stability or stasis, affective engagement more than detached contemplation, and generation of new artifacts (content, data, software tools) rather than reflection on already existing ones.

[8] There are several important points of view on this, but too little space to delve into all of them (Helberger et al., 2018; Napoli and Caplan, 2017; van Dijck et al., 2018).

10.4 Objectives and Ethics

Now that we have summarized Facebook as a largely automated, techno-commercial architecture collecting unimaginably vast, impenetrable, and non-overseeable user data in the purpose of guiding users in various behavioral directions, we will begin summarizing some of the ethical implications of this.

Of course, there are democratic upsides to the ways in which social media platforms offer a place for a wide range of ideas to be shared by broad arrays of groups and individuals (Daniels, 2014, p. 143), potentially beneficial to society at large. It is not my task in this chapter to repeat those connectionist claims by the corporations themselves, but rather to try to identify some possible drawbacks of this particular form of systemic openness, since it provides new means to manipulate, filter, and broadcast both truthful and purposely false information (Schou and Farkas, 2016, p. 37). This is the kind of epistemological challenge that would traditionally have been ameliorated by improving the abilities of, for example, writers, editors, publishers, broadcasters, and librarians to assess the quality of (intentionally published) information online. Gillespie (2018) argues that content moderation is at the definitional core of social media platforms. Given all the atrocities posted on such platforms, it is no longer controversial to argue that platforms need moderation. The question remains what type of responsibilities platforms should have and the difficulties of policing and normative decision-making. This particularly becomes apparent in boundary cases, when contexts differ and contradictions proliferate. Malicious actors misuse features like tagging and flagging in order to obstruct opponents and shape the conditions for discourse. Automated policy implementation cannot avoid challenges of definition and navigation in and among a variety of cultures, conflicting legal systems, and inflamed issues (Gillespie, 2018, p. 10).

These moderation challenges aside, whether the users make conventionally rational decisions or not is of little interest to Facebook's managers, as long as there are mechanisms that also channel these decisions to be expedient for the purposes of the system at large. Facebook is not in the service of creating reliable, valid, and representative information. Academic scholars often look for such things; but one needs to remember the many other sectors of society where endeavors of reliability, validity, and representativity are indeed secondary, compared to more expedient values like profitability, efficacy, and social control. The reason why Facebook's databases have to be reliant, workable, and the data therein adequately unambiguous is less due to claims to "objective knowledge" than due to claims to solid market valuation. Infrastructures like these are less about knowing than about defining and valuing (Cheney-Lippold, 2017; Srnicek, 2016; van Dijck et al., 2018).

Moreover, in ethical terms, Mirowski (2019) presents a rather heady argument that platforms like Facebook, being products of neoliberal market ideas, in practice operate as tools of hegemony. Neoliberal epistemology stipulates that, as individuals, people are sloppy, undependable cognitive agents, he argues. But if they come to engage as the constituent parts of collective markets, then order will emerge spontaneously since neoliberal doxa stipulates markets are the most excellent information processors imaginable.[9] The task of the platform design is *to make people behave in line with this theory of market behavior*. Consequently, "fake news" – defined as the specific phenomenon of stories purposely fabricated for clicks and revenue (Benkler et al., 2018; Farkas and Schou, 2019; Gillespie, 2018) – reverberates very well with core philosophical presuppositions of neoliberalism, Mirowski argues. Fake news distributes a set of fabrications, conveyed and validated by the market, that act to help keep the populace misinformed so that the true workings of the elite in-group can continue undisturbed. This was actually presented as a deliberate political strategy by neoliberal pioneers, for example, Strauss (1952), Mirowski points out. Similarly, Coase (1950) argued against the Reithian educational heritage of the BBC and instead stipulated that the market should predicate – and, as Mirowski puts it, *if the masses want trash, let the market give them trash*. In the contemporary era, we see how compatible this is with the information oversaturation doctrines stipulated by Vladimir Surkov (Pomerantsev, 2011) and, during the Trump era, implemented also by the Republican Party in the USA (Coppins, 2020).

In other words, ethics – as media and communications scholars commonly think of it[10] – are not easily applicable here. This is a new type of ethical and political Leviathan. Facebook as a commercial infrastructure is not an editorial one. Despite appearing to become more responsive to public concerns and moral outrages, I argue that the corporation has so far remained largely indifferent to what types of moralism, judgments, or arguments its users use the platform to put forward. For Facebook, the important thing is that users' future informational and physical environments are attuned "according to the predictions contained in the statistical body" (Rouvroy, 2013, p. 157). If we are to follow the platform logic at hand to its logical conclusion, it is also obvious that the issue of whether to "confront 'subjects' as moral agents" (ibid.) in any conventional sense is, for Facebook as a corporation, not that important, unless it has a bearing on user retention (rates of churn), user engagement (rates of activity), and/or their exposure to ads (rates of ad interaction).

[9] The key mechanism here is catallaxy, as stipulated by (Hayek, 2012, p. 107–132).

[10] I.e., the way ethics tends to be conceptualized as "media ethics" or the "professional ethics" of journalism.

10.5 Conclusion: Perspective-Making on, of, and through Facebook

Where does this all leave us, then, in terms of everyday sensemaking and mundane practices among media users of making meaning in the world? Communication as an academic field has always been attuned to the very construction of arguments about what "counts" as relevant information, opinion, and choice (Cheney et al., 2011, p. 1), and a key heuristic is provided throughout this book, namely the diffractive, heterogeneous nature of epistemology and ethical reasoning, in relation to communicative practices. There are diverse regimes of justification (Boltanski and Thévenot, 2006), assigning weight to rather different things, differently, according to different (sometimes incommensurable) scales of value. Any endeavor to trace different communicative repertoires or modes of argumentation on Facebook would have to take into consideration the "shared ethical norms along with the irreducible differences between diverse cultural traditions and communicative preferences" (Ess, 2011, p. 205) and, when applicable, "disciplinary enclosures of law, trade, and communication technologies, each with their own normative codes" (D'Souza, 2011, p. 475), as well as being alert to possible disjunctures between theoretical inquiries and popular, applied communications ethics (Cheney et al., 2011, p. 4). Moreover, as other chapters in this volume show, there are contextual forces operating at different scales: Particular group dynamics, micro and macro, where minute decisions in small groups interact with the larger system in various ways.

What can be learnt from my analysis of the key dimensions that play a role in the circulation of information through Facebook, and how should researchers proceed when evaluating information on the platform? Since all social action on platforms like Facebook is so highly mediated and shaped by its interface design (i.e., the "choice environment" that it constitutes), any theory of the ways in which perception and sensemaking are taking place on and through Facebook has to consider its abductive nature: Users "make do" with the indications of social action, of various modes of being, and of distinctions of worth presented to them. Interestingly, this is a mode of sensemaking which applies also to the corporation's own analysts, who would also have to "make do" with what the signals gathered through the interface tells them. It also applies to researchers trying to glean knowledge from the mediated data. It is obvious that what is presented through such interfaces can be wildly misleading. Perhaps a good metaphor for these interfaces would be that they in many ways constitute veils, through which minute signals pass, indicating an imagined reality on the other side.

One way to compensate for this veil of ignorance and speculation, as one could call it, and the communicative reductionism inherent to all digitization, would be to employ qualitative methods, as a complement to one's sparse, signals-based intelligence – but even so, any reflexive account of experiences would be based on experiences that would originally have been equally interface-based and abductive. Arguably, the only real antidote to speculation, reductionism, and ignorance is *contextual knowledge*: the ability to reasonably judge and reflect on the validity and reliability of what one has seen.

Moreover, since the communication that circulates through the Facebook infrastructure is of such a textual kind, comprising of static signs (texts, pictures, videos) it is bound to be woefully prescriptive, compared to oral culture, which is not comprised by symbolic (thus prescriptive) artifacts. Facebook's sorting algorithms, elevating certain content, help to amplify this prescriptive power. Moreover, dynamic, numerical elements (e.g., "like" and "share" counts) indicate popularity and salience (hereby indicating additional social and cultural worth) where there might actually be none. Facebook's global infrastructural significance prescribes institutional actors to align their organizations to be interoperable with Facebook. By extension, material aspects like these help compounding normative weight, not only as to *what* is published through Facebook, but the fact, in itself, *that* it is published, and *the ways in which* its communicative modes and schemata work, often in stark contrast with conventional journalistic norms of vetting and/or editing.

Where does this lead the future of research on social media platforms? Ultimately, if Facebook as a techno-commercial architecture allows for openness, it is a very mandated form of openness: Certain ways of expressing oneself are favored, certain algorithmic feedback loops are generated, depending on what is deemed by Facebook the corporation to be expedient to circulate and enhance further in each respective user experience. The categorization-at-a-distance (Cheney-Lippold, 2017) that the data analysts perform has a real bearing on how individuals get to be labeled and categorized and, in effect, different forms of social sorting that play into the ways in which people conceive of themselves as persons and the ways in which they perceive each other. None of this is made clearer by Facebook's strong policy of secrecy and discretion, not even allowing independent researchers access to the administrative back-end decisions shaping these social conditions. "For every epistemological challenge the seemingly black-boxed algorithm poses, another productive methodological route may open" (Bucher, 2018, p. 64).

In many ways, my own inquiry has echoed that of Bucher's. Thankfully, she argues (pp. 64–65), there is reason not to despair. Do not let the opacity of the infrastructure set epistemological limits that impinge knowledge-making. The opacity is never *total*, as I understand her, and by tracing, outlining the nature of the black box itself, observing and describing its limits, some knowledge about it will be generated, after all. Seek out ways to gain knowledge despite these partial opacities; triangulating observational data from sources adjacent to the black box will likely help. Ultimately, since no data are entirely objective, the data assumed to reside inside the black box should never be expected to be objective nor absolute.

References

Agre, P. 1994. Surveillance and capture: Two models of privacy. *The Information Society*, **10**(2), 101–127.

Andersson Schwarz, J. 2016. Public service broadcasting and data-driven personalization: A view from Sweden. *Television & New Media*, **17**(2), 124–141.

Andersson Schwarz, J. 2018. Umwelt and individuation: digital signals and technical being. Pages 61–80 of: Lagerkvist, A. (ed), *Digital Existence*. London & New York: Routledge.

Barthes, R. 1964. Rhetoric of the image. Pages 32–51 of: *Image–music–text (1977)*. Sel. and Trans. Stephen Heath. New York: Hill and Wang.

Benkler, Y., Faris, R., and Roberts, H. 2018. *Network propaganda: Manipulation, disinformation, and radicalization in American politics*. Oxford: Oxford University Press.

Boltanski, L. and Thévenot, L. 2006. *On justification: Economies of worth*. Trans. C. Porter. Princeton, NJ, & Oxford: Princeton University Press.

Bucher, T. 2018. *If...Then: Algorithmic power and politics*. Oxford & New York, NY: Oxford University Press.

Caplan, R. and Boyd, D. 2018. Isomorphism through algorithms: Institutional dependencies in the case of Facebook. *Big Data & Society*, **5**, 1.

Cheney, G., Munshi, D., May, S., and Ortiz, E. 2011. Encountering communication ethics in the contemporary world: Principles, people, and contexts. Pages 1–14 of: Cheney, G., May, S., and Munshi, D. (eds), *Cheney. Handbook of communication ethics*. London & New York: Routledge.

Cheney-Lippold, J. 2017. *We are data: Algorithms and the making of our digital selves*. New York, NY: New York University Press.

Coase, R. 1950. *British broadcasting: A study in monopoly*. London: London School of Economics.

Coppins, M. 2020. The billion-dollar disinformation campaign to reelect the president. *The Atlantic*, March 10. https://www.theatlantic.com/magazine/archive/2020/03/the-2020-disinformation-war/605530/.

Daniels, J. 2014. From crisis pregnancy centers to Teenbreaks.com: Anti-abortion activism's use of cloaked websites. Pages 140–154 of: McCaughey, M. (ed), *Cyberactivism on the Participatory Web*. London & New York, NY: Routledge.

D'Souza, R. 2011. When unreason masquerades as reason: Can law regulate trade and networked communication ethically? Pages 475–493 of: Cheney, G., May, S., and Munshi, D. (eds), *Handbook of communication ethic*. London & New York: Routledge.

Ess, C. 2011. Ethical dimensions of new technology/media. Pages 204–220 of: Cheney, G., May, S., and Munshi, D. (eds), *Handbook of communication ethics*. London & New York: Routledge.

Farkas, J. and Schou, J. 2019. *Post-truth, fake news and democracy: Mapping the politics of falsehood*. London & New York, NY: Routledge.

Gerken, M. 2017. *On folk epistemology: How we think and talk about knowledge*. Oxford: Oxford University Press.

Gibson, J. J. 1979. *The ecological approach to visual perception*. New York, NY & London: Psychology Press.

Gillespie, T. 2014. The relevance of algorithms. Pages 167–194 of: Gillespie, T., Boczkowski, P. J., and Foot, K. A. (eds), *Media technologies: Essays on communication, materiality, and society*. Cambridge MA & London: MIT Press.

Gillespie, T. 2018. *Custodians of the Internet: Platforms, content moderation, and the hidden decisions that shape social media*. New Haven, CT: Yale University Press.

Hacking, I. 1995. The looping effect of human kinds. Pages 351–383 of: Sperber, D., Premack, D., and Premack, A. J. (eds), *Causal cognition: A multidisciplinary debate*. Oxford: Clarendon Press.

Hacking, I. 2004. *Historical ontology*. Cambridge, MA: Harvard University Press.

Hall, S. 1980. Encoding/decoding. Pages 128–138 of: Hall, S., Hobson, D., Lowe, A., and Willis, P. (eds), *Culture, media, language: Working papers in cultural studies*. London: Hutchinson.

Hayek, F. August. 2012. *Law, legislation and liberty, volume 2: The mirage of social justice*. Vol. 2. Chicago, IL: University of Chicago Press.

Helberger, N., Pierson, J., and Poell, T. 2018. Governing online platforms: From contested to cooperative responsibility. *The Information Society*, **34**(1), 1–14.

Helmond, A. 2015. The platformization of the Web: Making Web data platform ready. *Social Media + Society*, **1**(2), 1–11.

Jasanoff, S. 2015. Future imperfect: Science, technology, and the imaginations of modernity. In: Jasanoff, S., and Kim, S. H. (eds), *Dreamscapes of modernity: Sociotechnical imaginaries and the fabrication of power*. Chicago, IL: University of Chicago Press, 1–33.

Lessig, L. 1999. *Code and other laws of cyberspace*. New York, NY: Basic Books.

Magnani, L. 2009. *Abductive cognition: The epistemological and eco-cognitive dimensions of hypothetical reasoning*. Berlin: Springer.

Mansell, R. 2012. *Imagining the Internet*. Oxford: Oxford University Press.

McCulloch, W. S. 1965. *Embodiments of mind*. Cambridge, MA: MIT Press.

Mirowski, P. 2019. Hell is truth seen too late. boundary 2. *Boundary*, **46**(1), 1–53.

Napoli, P. and Caplan, R. 2017. Why media companies insist they're not media companies, why they're wrong, and why it matters. *First Monday*, **22**(5).

O'Callaghan, D., Greene, D., Conway, M., Carthy, J., and Cunningham, P. 2015. Down the (white) rabbit hole: The extreme right and online recommender systems. *Social Science Computer Review*, **33**(4), 459–478.

Parisi, L. 2017. Computational logic and ecological rationality. Pages 75–99 of: Hörl, E. (ed), *General ecology: The new ecological paradigm*. London: Bloomsbury Academic.

Parks, L. 2012. Technostruggles and the satellite dish: A populist approach to infrastructure. Pages 64–86 of: Bolin, G. (ed), *Cultural technologies: The shaping of culture in media and society*. New York, NY: Routledge.

Plantin, J.-C., Lagoze, C., Edwards, P. N., and Sandvig, C. 2018. Infrastructure studies meet platform studies in the age of Google and Facebook. *New Media & Society*, **20**(1), 293–310.

Pomerantsev, P. 2011. Putin's Rasputin. *London review of books*, **33**(20), 3–6.

Rouvroy, A. 2013. The end(s) of critique: Data behaviourism versus due process. Pages 143–168 of: de Vries, K. and Hildebrandt, M. (eds), *Privacy, due process, and the computational turn*. London and New York: Routledge.

Schou, J. and Farkas, J. 2016. Algorithms, interfaces, and the circulation of information: Interrogating the epistemological challenges of facebook. *KOME – An International Journal of Pure Communication Inquiry*, **4**(1), 36–49.

Slack, D. and Wise, J. M. 2002. Cultural studies and technology. Pages 485–501 of: Lievrouw, L. and Livingstone, S. (eds), *The handbook of new media*. London, Thousand Oaks, CA & New Delhi: Sage.

Srnicek, N. 2016. *Platform capitalism*. Cambridge & Malden, MA: Polity.

Stewart, J. (2011). A contribution to ethical theory and praxis. In Cheney, G.; S. May; D. Munshi (Eds.) *Handbook of communication ethics*. London & New York: Routledge. 15–30.

Strauss, L. 1952. *Persecution and the art of writing*. Chicago: University of Chicago Press.

Sugarman, J. 2014. Neo-Foucaultian approaches to critical inquiry in the psychology of education. Pages 53–69 of: Corcoran, T. (ed), *Psychology in education: critical theory~practice*. Rotterdam, Netherlands: SensePublishers.

Sylvain, O. 2018. Intermediary design duties. *Connecticut Law Review*, **50**(1), 204–274.

Thaler, R. H. and Sunstein, C. R. 2008. *Nudge: Improving decisions about health, wealth and happiness*. London: Yale University Press.

van Dijck, J. 2013. *The culture of connectivity: A critical history of social media*. Oxford: Oxford University Press.

van Dijck, J., Poell, T., and Waal, M. De. 2018. *The platform society: Public values in a connective world*. Oxford: Oxford University Press.

von Foerster, H. 2003. *Understanding understanding: Essays on cybernetics and cognition*. Berlin: Springer.

Webster, J. 2014. *The marketplace of attention: How audiences take shape in a digital age*. Cambridge, MA & London: MIT Press.

Whitehead, A. N. 1929. *The function of reason*. Boston: Beacon Press.

Wu, T. 2016. *The attention merchants: The epic scramble to get inside our heads*. New York, NY: Knopf.

11

Content, Form, and Reception: Perspectives from Digital Media Data

Christina Neumayer

11.1 Introduction

This chapter explores how communication on digital platforms takes place by analytically separating content, form, and reception. With illustrative examples from activists' platform communication, it discusses how these classic concepts of communication theory can be used to identify and analyze perspectives communicated on the Web. Content, form, and reception have always been central concepts of communication studies. The focus on these concepts goes back to Shannon and Weaver's linear communication model where a sender sends a message to a receiver. This model was later revised by Jacobsen (among others), adding context and channel as well as a feedback loop (for an overview, see McQuail, 1987). These mass media–based models, however, tended to reduce the complexity of mediated communication, which is further complicated today by the sociocultural and techno-commercial perspectives imbricated in online information, the topic of this book. Analytically separating content, form, and reception makes it possible to understand which perspectives are over- and underemphasized in platform data.

This chapter models the factors that shape particular content-level perspectives in online communication, as well as the form in which these perspectives are delivered and received. Information is often communicated through text, but comprehensive understanding of the perspectives embedded in a message also requires analysis of the message's audiovisual elements, which are becoming increasingly important on today's Web. The context in which information is shared is also crucial. The perspectives embedded within messages communicated via online platforms are shaped through processes of reception, that is, by commenting, liking, sharing, voting, and flagging.

Today, researchers and analysts often discern assumptions about communication from digital media data. Returning to classic models of communication theory reminds us that we need to understand digital media data (in the same way as communication) as a process. Social scientists have a long tradition of understanding data, where they come from, and what they actually represent (Mattoni and Pavan, 2018). To understand the perspectives communicated in such data, we have to start from the assumption that all data are made, as they are "cooked" (Gitelman, 2013) by underlying sociotechnical processes. If we understand data as made and thus processual and relational, the perspectives we discern from such data are also never simply there waiting for us to be discovered. Perspectives in online communication are transformed into analyzable data based on which we draw conclusions. With this in mind, we argue that digital media data are never just there or simply visible. There are sociotechnical practices at play while these data are made and made visible (see Neumayer et al., 2021).

In the following, we trace content, form, and reception of perspectives in online communication that are discerned from digital media data. This process starts when data are (a) created as activists interact with platforms in the specific context of political protest (data creation phase), (b) datafied by digital media platforms (datafication phase), and (c) retrieved and analyzed by analysts and their methods (data retrieval and analysis phase). For the purpose of clarity, we analytically divide between these three phases and between content, form, and reception as key aspects of online communication. Before we focus on how this process takes place, we will briefly discuss the particularities of political protest.

11.2 A Brief Note on Perspectives from Political Protest

Digital media data have grown to become a large and rich evidence base for political protest, but much data remains out of sight, rendering certain perspectives invisible or underrepresented. Simultaneously, the research methods we employ turn our attention to specific factors of content (e.g., content such as violence or humor), form (e.g., network structures, discernible as text), and reception (e.g., retweets and replies) while others remain invisible. This research is based on the assumption that digital traces can provide deep insight into perspectives from political protest. Yet, as data repositories of social movements' perspectives, they grant only a limited perspective on contemporary protests.

Political protest takes place between a need for producing visibility (to mobilize) but simultaneously remains invisible (when acting in civil disobedience) (Neumayer and Stald, 2014). Protest does not happen in "flat" and evenly distributed opportunities. Imbalances in power in terms of access to visibility have always complicated the relationship between activists and media. Gamson and Wolfsfeld (1993) described movements and media as interacting systems engaged in a struggle over meaning of protest events as a legitimate form of action to express legitimate grievances. They stress a power imbalance and dependency that favors media at the expense of social movements, given that gaining voice depended entirely on representation by journalists. Movements must struggle to establish their standing if mass media are to convey their message – rather than distort, mistranslate, or ignore it entirely (Gamson and Wolfsfeld, 1993, p. 117). In parallel, certain ideas, values, and languages are welcomed while others (such as radical voices) are less popular or rendered invisible by media norms and practices.

In digital media, traditional gatekeepers (such as journalists and authorities) are still there, but the dynamics are complicated by algorithms (such as trending topics), business models, services, and policies of digital media corporations. In the following, we outline how perspectives of and from protest are discerned from digital media data of online communication during the data creation, datafication, and data retrieval and analysis phase. We analytically separate content, form, and reception as aspects of online communication. Using political protest as an illustrative case, we identify how each aspect takes shape when discerning information from digital media data.

11.3 Content

Analyzing the content of classic mass media follows well-established methods such as content analysis, discourse analysis, and frame analysis; the object of study is well defined such as headlines and/or text body and TV news; and the question of visibility can be measured by number of viewers or readers. This becomes more complicated when we base our findings on digital media data.

11.3.1 Data Creation Phase

Activists use increasingly sophisticated techniques to mobilize, organize, spread their message, and tell their story through digital media, in other words, to become visible. While we often talk about visibility as something that is positive while the invisible is excluded, underrepresented, and unseen, visibility can also lead to social constraints through surveillance and control. Techno-commercial digital media platforms can amplify the visibility of social movements but also shut down opportunities by surveillance and control (Neumayer and Stald, 2014) or undermine the voicing of critique altogether (Uldam, 2018). The data creation phase is mainly concerned with the user handling the technologies and economic models underlying digital media platforms. With the entanglement between media technologies and activism (see e.g., Galis and Neumayer, 2016) and corporate digital media having become "algorithmic mass media" (Leistert, 2015) censorship through algorithms can become a normalization and standardization tool for activists' communicative action. Activists can employ strategies to creatively tamper with data or use bots, conceal, obfuscate and deceit, avoid online traces (Brunton and Nissenbaum, 2015; Youmans and York, 2012), or deradicalize content (Neumayer, 2016). All these tactics ranging from producing visibility to avoidance, intentional silencing, and obfuscation complicate the question of which perspectives become visible in content of online communication based on digital media data.

11.3.2 Datafication Phase

The datafication phase is concerned with the process of turning the content activists produce into data and what that in return does to such content. Invisible algorithms and the techno-commercial materiality of digital media may increase the visibility but also traceability of activist communication (see Mortensen et al., 2019). Social media corporations may remove content that allegedly breaks terms of service and disturbs processes of community formation (Youmans and York, 2012). They can decide whether to make pseudonyms possible, render traces ephemeral or permanent (Tufekci, 2017). This can mean a reproduction of classic bias in reporting from political protest such as focus on violent forms of action and escalation (Poell and Borra, 2011) as well as content that works well in digital media environments such as humor (Jensen et al., 2020). New forms of information diffusion and representation within the sociotechnical structure of the different platforms may privilege spectacular, violent, and viral images over content that illuminates wider societal and political issues or undermine the voicing of political critique altogether (Neumayer and Rossi, 2018).

11.3.3 Data Retrieval and Analysis Phase

There is a long tradition of studying bias in media content, and we usually differentiate between selection bias (as media cover a small selection of protests; see McCarthy et al. (1996)) and description bias (the frames, grievances, and demands media pay attention to; see Smith et al. (2001)). While these forms of bias are still there, inquiries of bias have become more complex with digital media data (Hargittai, 2020) for several reasons: No clear reference for bias introduced by institutions (Shoemaker and Reese, 2014), content generated by users (Walther and Jang, 2012), opaqueness of algorithmic processes

(Boyd and Crawford, 2012), and the idea that digital media data are data per se and not content produced by people (journalists) and then turned into data (by researchers). Moreover, while content produced within media institutions is usually embedded in a well-defined context, digital media content is often de-contextualized. Even though content might appear in a specific hashtag, it might not necessarily be related to the protest event in question. Collecting and analyzing such data as a representation of political protest focuses our attention on content that is overemphasized in digital media, accessible, and analyzable. Doing so renders any content that is not accessible and traceable invisible (e.g., content on closed platforms such as WhatsApp).

11.4 Form

Perspectives on the Web discerned from digital data depend on: the form in which people communicate through digital media platforms; in which form the data are archived and sorted; and in which form they can be accessed, collected, and analyzed. This shapes what perspectives from political protest become visible in digital data (Neumayer and Struthers, 2019).

11.4.1 Data Creation Phase

To express their political grievances, activists may need to adjust to forms of communication that are privileged on online platforms such as memes and generally visual content (Blaagaard et al., 2017), and short slogans or personalized stories that travel well through social media platforms (Bennett and Segerberg, 2012). They create data (often without being aware of it) in a form that complies with the social media logic (van Dijck and Poell, 2013). Social movements adjust their media practices to the immediacy of digital platforms, which demand instantaneous and continuous forms of communication by activists to compete for visibility with the constant noise produced on digital media platforms (Barassi, 2015). Packaging their political grievances into constant, personalized, short, visual, and catchy forms that gain attention in digital platforms can hollow out political activism. Grievances and demands may be reduced to communication that fits into a certain form such as a tweet, image, hashtag, post, comment, meme, GIF (Graphics Interchange Format), or online video.

11.4.2 Datafication Phase

Certain forms of communication are privileged by the techno-commercial infrastructure of digital media and travel well through online platforms. As they travel through the Web, they might become decontextualized and further divert the attention from political critique. This can, for example, be the case with memes that start from a political protest but then are photoshopped into various contexts mainly for the purpose of creativity and entertainment (see Bayerl and Stoynov, 2016; Jensen et al., 2020). Political resistance on the Web is often based on ephemeral and unstable forms of collectivity by people sharing their personalized stories on social media which are then filtered and sorted by hashtags or trending topics (see Poell and van Dijck, 2015). The sorting, organizing, and filtering of short and catchy slogans, images, and other visual forms such as memes into more or less coherent narratives by digital media platforms can further distort and undermine collective action. These can be connected by links, retweets, tags, favors, likes, profiles, or other digital forms but also by physical objects in images. The Occupy campsites as an expression of economic inequality (Feigenbaum, 2014), the umbrella (initially used to protect activists from tear gas) as a symbol of resistance against authorities in Hong Kong (see Chan, 2014) are examples of such objects communicating resistance and distinguishing movements from one another. As objects of resistance (the umbrella or the tent) travel through digital media as images – mediated and remediated, appropriated, memefied, and personalized – they become symbols of resistance and forms of activism in their own rights. They are classified through tags and feeds and archived, and we inscribe meaning to images and texts based on such tags (e.g., hashtags, automated image tags).

11.4.3 Data Retrieval and Analysis Phase

Digital media data, accessed through APIs provided by large corporations, have been used as a basis for statements of fact, despite these data being likely to be incomplete because they are structured through algorithms that remain opaque (Neumayer and Rossi, 2016; Neumayer et al., 2021). The advantages of using computational methods for the study of political contention are that we can draw on unprecedented sources of social data in various forms to understand dynamics of collective action (Margetts et al., 2015). Yet, the "regimes of access" (Burgess and Bruns, 2012) of digital media companies regulate the form in which researchers can collect and use data. Such conditions direct scholarly attention toward relatively open and text-based platforms (e.g., Twitter) and collection strategies that follow their functionalities (e.g., hashtags, profiles). A comprehensive understanding of perspectives requires analysis of the message's audiovisual element, but there are advanced methods for text-based analysis, while computational analysis of visual content lags behind (Bennett, 2019; Neumayer and Rossi, 2016). Certain methods make perspectives visible in a particular form that determines what researchers are able to analyze. Relational data, for example, has been visualized as networks that are not "naturally" there but made visual in a particular form, while lists would be analyzed with methods such as natural language processing (Neumayer et al., 2021). Similarly, Facebook's decision in 2018, for example, to restrict various forms of data for researchers led to fundamental changes in the design of research projects aimed at understanding social complexity based on an analysis of digital media data (see Walker et al., 2019). While the form in which perspectives on the Web occur is often treated as a given, digital platforms shape the form in which we understand and analyze perspectives on the Web.

11.5 Reception

Reception is being complicated on the Web, as the audience is not clear. We often rely on modes of reception as a measurement of visibility when relying on digital data. The perspectives embedded within messages communicated via online platforms are thus shaped through processes of reception, that is, by commenting, liking, sharing, voting, and flagging. This includes the context in which information is shared and received.

11.5.1 Data Creation Phase

Individuals contribute to a cause by sharing, liking, and other forms of expressing solidarity through digital media, but by doing so, they also adhere to the digital media logic and give up control over their data in their struggle over visibility (Poell and van Dijck, 2015). Reception is complicated, as it may change the content of a message (such as a repost or retweet including a message that changes the original meaning of the post). Simultaneously, antagonistic actors such as police successfully employ these forms of reception to interrupt a narrative of protest events by interfering in it through commenting (Neumayer and Rossi, 2018). This might obscure activists' carefully planned digital media tactics. Visibility is thus inherently unstable, as narratives, symbols, and images can move from the visible into the invisible and vice versa. Through modes of reception perspectives can be changed, messages can be emphasized or conversely rendered invisible. At the same time, these modes of reception are visibly attached to a person's identity when digital platforms include profiles with real names. People create data about themselves by using modes of reception but at the same time modify data that is already there. Doing so, the modes of reception might also divert our attention to uncritical humorous perspectives, which we are more likely to visibly support through retweets, likes, and shares.

While images of violence, riots, and police brutality might be seen, we may be less inclined to like or favor them. Data on platforms is thus never stable but constantly created through new relations, processes, and data created by modes of reception.

11.5.2 Datafication Phase

Parallel to potential threats to activists' agency through the collection of data, the likes, shares, follows, and favors of particular posts, images, and messages also lead to overexposure of certain messages and concerns while others remain invisible. Digital media environments privilege information that has been visibly received by such expressions of support. This, however, requires a user account and despite certain platforms (such as Reddit) allowing for pseudonyms (see Triggs et al., 2021), the number of likes, favors, and upvotes, is not necessarily indicative of the actual reception of information. Digital media's "technological features, such as 'retweeting', liking, following, and 'friending', as well as algorithmic selection mechanisms [...] privilege particular types of content" (Poell and van Dijck, 2015, p. 528), as they do collect more likes and shares that in return can be turned into data feeding into digital media corporations' business models. Automated and manual forms of curation and moderation are informed by business models focused on maximizing user engagement without alienating the overall user population, advertisers, and other commercial partners (van Dijck et al., 2018). Activists' communication in digital media might not always be coherent with such values and business models (as they might not visibly show support through likes and favors), and as a consequence, their perspectives may be rendered invisible.

11.5.3 Data Retrieval and Analysis Phase

Digital media data might be collectively produced in an Instagram/Twitter hashtag, a Facebook page, and a YouTube channel. However, "activists have control over neither the social media architectures through which they communicate nor the data they collectively produce through Twitter hashtags and Facebook pages and groups" (Poell, 2015, p. 199). What happens with the data they produce is only visible to digital media corporations, and the data they share can be used in unintended contexts. Stefania Milan described the visibility of collective action in digital media as "the digital embodiment and online presence of individuals and groups and their associated meanings, which are (and need to be) constantly negotiated, reinvigorated, and updated" (Milan, 2015, p. 7). "Semantic units" (Milan, 2015) composed of algorithmically assorted datafied emotions, images, and messages are constantly reproduced and modified, producing visibility for an unstable user-generated narrative. Collective identity, actions, and grievances are thus not a constant but constantly in the making, as they have an afterlife in digital media but are then studied as one phenomenon and as they are connected through modes of reception. These modes of reception then are used as measurements (such as number of retweets and likes) of visibility and determine how and what analysts pay attention to (such as focus on hashtags that can produce large retweet networks; large number of posts that require automated text analysis).

11.6 Perspectives from Digital Data

To understand perspectives from the Web based on digital data, we need to trace how such data are made and processed. Analytically dividing that process into content, form, and reception and by understanding digital data as a process including the three phases of data creation, datafication, and retrieval and analysis allows us to understand what becomes overemphasized and what remains invisible (see Figure 11.1). While this chapter does not provide a complete analysis of such processes, it rather invites us to understand how digital media data are made and to analytically separate between content, form, and reception. Taking the processual understanding from communication theory and returning to classic models of communication encourages us to think about what digital data actually represent. Distancing ourselves from the assumption that all digital data are visible and representative of all perspectives on the Web can eventually help develop a more reflective field of research based on digital media data. While a large number of perspectives from political protest may be visible, traceable, and analyzable on the Web, they are shaped by the underlying sociocultural and techno-commercial processes taking place on digital media platforms. Awareness of these processes can enhance our understanding of the perspectives as they not only appear on the Web but are shaped by a variety of sociotechnical actors.

While the "pictures in our heads" (as famously coined by Lippman (1922)) produced by mass media never represented the world as it is, we have a relatively good understanding of who makes these pictures and the power dynamics at play. The representational power in digital media, however, also lies in the "pictures of our actions made of bits, stored in databases, available to algorithms" (Turner, 2014, p. 254), which complicates such processes. It is necessary to understand that not all digital data is visible and that knowledge of that which is invisible can enhance our understanding of perspectives discernible in

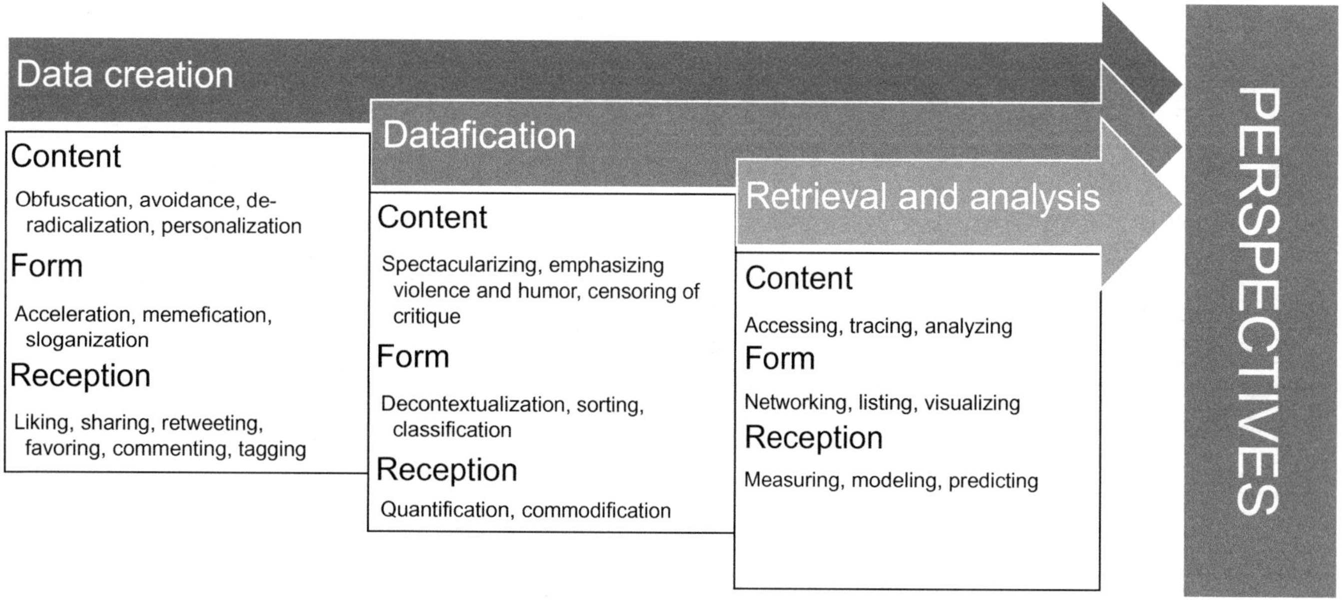

Figure 11.1 Three phases of making digital media data including processes that shape content, form, and reception.

social media data from political protest. A Perspective Web that renders all the invisibilities visible would help us to discern more adequate and representative results based on digital media data.

In a discussion of the representation of political protest, we have argued that turning our attention to the invisible and under-represented might tell us something about power relations in activists' struggles. Yet, a Perspective Web that renders all interactions on the Web visible would also render visible perspectives that were intentionally made invisible. This becomes particularly relevant for activists who act in civil disobedience and intentionally silence their interactions. Connecting invisible data points to digital methods requires us to understand invisibilities per se while at the same time respecting the invisibilities and silences intentionally created by activists who are embedded in an imbalance of power. While creating visibility for activists' grievances and injustice is important, a Perspective Web might, in some situations, also put people at risk. However, turning our attention to data invisibilities with a Perspective Web might help to understand how the visible distracts from invisibilities, how invisibilities can reveal both intentional silences and power relations in activists' struggles for attention, and how the sociotechnical materiality of social media corporations influences both political protest and our assumptions concerning the ways in which perspectives from political protest are discernible in digital media data.

References

Barassi, V. 2015. Social media, immediacy and the time for democracy: Critical reflections on social media as "temporalizing practices." Pages 73–88 of: Dencik, L. and Leistert, O. (eds), *Critical perspectives on social media and protest: Between control and emancipation*. London, New York: Rowman & Littlefield.

Bayerl, P. S. and Stoynov, L. 2016. Revenge by photoshop: Memefying police acts in the public dialogue about injustice. *New Media & Society*, **18**(6), 1006–1026.

Bennett, W. L. and Segerberg, A. 2012. The logic of connective action: Digital media and the personalization of contentious politics. *Information, Communication & Society*, **15**(5), 739–768.

Blaagaard, B., Mortensen, M., and Neumayer, C. 2017. Digital images and globalized conflict. *Media, Culture & Society*, **39**(8), 1111–1121.

Boyd, D., and Crawford, K. 2012. Critical questions for big data: Provocations for a cultural, technological, and scholarly phenomenon. *Information, Communication & Society*, **15**(5), 662–679.

Brunton, F. and Nissenbaum, H. 2015. *Obfuscation: A user's guide for privacy and protest*. Cambridge, MA: MIT Press.

Burgess, J. and Bruns, A. 2012. Twitter archives and the challenges of "Big Social Data" for media and communication research. *M/C Journal*, **15**(5).

Chan, J. 2014. Hong Kong's umbrella movement. *The round table*, **103**(6), 571–580.
Feigenbaum, A. 2014. Resistant matters: Tents, tear gas and the "other media" of Occupy. *Communication and Critical/Cultural Studies*, **11**(1), 15–24.
Galis, V. and Neumayer, C. 2016. Laying claim to social media by activists: a cyber-material détournement. *Social Media + Society*, **2**, 3.
Gamson, W. A. and Wolfsfeld, G. 1993. Movements and media as interacting systems. *The Annals of the American Academy of Political and Social Science*, **528**(1), 114–125.
Gitelman, L. (ed). 2013. Hargittai *Raw data is an oxymoron*. Cambridge, MA: MIT press.
Hargittai, E. 2020. Potential biases in big data: Omitted voices on social media. *Social Science Computer Review*, **38**(1), 10–24.
Jensen, M. S., Neumayer, C., and Rossi, L. 2020. "Brussels will land on its feet like a cat": Motivations for memefying #Brusselslockdown. *Information, Communication & Society*, **23**(1), 59–75.
Leistert, O. 2015. The revolution will not be liked: On the systematic constrains of corporate social media platforms for protest. Page 35–52 of: Dencik, L. and Leistert, O. (eds), *Critical perspectives on social media and protest: Between control and emancipation*. London: Rowman & Littlefield.
Lippman, W. 1922. *Public opinion*. New York: Macmillan.
Margetts, H., John, P., Hale, S., and Yasseri, T. 2015. *Political turbulence: How social media shape collective action*. Princeton, NJ: Princeton University Press.
Mattoni, A. and Pavan, E. 2018. Politics, participation and big data. *Introductory Reflections on the Ontological, Epistemological, and Methodological Aspects of a Complex Relationship*, **11**(2), 313–331.
McCarthy, J. D., McPhail, C., and Smith, J. 1996. Images of protest: Dimensions of selection bias in media coverage of Washington demonstrations, 1982 and 1991. *American Sociological Review*, **61**(3), 478–499.
McQuail, D. 1987. *Mass communication theory: An introduction (2nd ed.)*. Thousand Oaks, CA, US: Sage.
Milan, S. 2015. When algorithms shape collective action: Social media and the dynamics of cloud protesting. *Social Media + Society*, **1**, 2.
Mortensen, M., Neumayer, C., and Poell, T (eds). 2019. *Social media materialities and protest: Critical reflections*. London: Routledge.
Neumayer, C. 2016. Nationalist and anti-fascist movements in social media. Pages 296–308 of: A., Enli, G., Skogerbø, E., Larsson, O., A., and Christensen, C. (eds), *Bruns. The Routledge companion to social media and politics*. London: Routledge.
Neumayer, C. and Rossi, L. 2016. 15 years of protest and media technologies scholarship: A sociotechnical timeline. *Social Media + Society*, **2**, 3.
Neumayer, C. and Rossi, L. 2018. Images of protest in social media: Struggle over visibility and visual narratives. *New Media & Society*, **20**(11), 4293–4310.
Neumayer, C. and Stald, G. 2014. The mobile phone in street protest: Texting, tweeting, tracking, and tracing. *Mobile Media & Communication*, **2**(2), 117–133.
Neumayer, C. and Struthers, D. M. 2019. Social media as activist archives. Pages 86–98 of: Mortensen, M., Neumayer, C., and Poell, T. (eds), *Social media materialities and Protest: Critical reflection*. London: Routledge.

Neumayer, C., Rossi, L., and Struthers, D. M. 2021. Invisible data: A framework for understanding visibility processes in social media data. *Social Media+ Society*, **7**(1), 2056305120984472.

Poell, T. and Borra, E. 2011. Twitter, YouTube, and Flickr as platforms of alternative journalism: The social media account of the 2010 Toronto G20 protests. *Journalism*, **13**(6), 695–713.

Poell, T. and van Dijck, J. 2015. Social media and activist communication. Page 527–537 of: Atton, C (ed), *The Routledge companion to alternative and community media*. London, England: Routledge.

Poell, T. 2015. Social media activism and state censorship. *Social media, politics and the state: Protests, revolutions, riots, crime and policing in an age of Facebook, Twitter and YouTube*, 189–206.

Segerberg, A., and Bennett, L. 2019. Afterword: Lessons and puzzles in studying social media materialities and protest. Pages 156–162 of: Mortensen, M., Neumayer C., and Poell, T. (eds), *Social media materialities and protest: Critical reflections*. London: Routledge.

Shoemaker, P. and Reese, S. D. 2014. *Mediating the message in the 21st century*. New York: Routledge.

Smith, J., McCarthy, J. D., McPhail, C., and Augustyn, B. 2001. From protest to agenda building: Description bias in media coverage of protest events in Washington, DC. *Social Forces*, **79**(4), 1397–1423.

Triggs, A. H., Møller, K., and Neumayer, C. 2021. Context collapse and anonymity among queer Reddit users. *New Media & Society*, **23**(1), 5–21.

Tufekci, Z. 2017. *Twitter and teargas*. New Haven & London: Yale University Press.

Turner, F. 2014. The world outside and the pictures in our networks. Page 251–260 of: T. Gillespie, P. J. Boczkowski, and Foot, K. A. (eds), *Media technologies: Essays on communication, materiality, and society*. Cambridge, MA: MIT Press.

Uldam, J. 2018. Social media visibility: challenges to activism. *Media, Culture & Society*, **40**(1), 41–58.

van Dijck, J. and Poell, T. 2013. Understanding social media logic. *Media and Communication*, **1**(1), 2–14.

van Dijck, J., Poell, T., and de Waal, M. 2018. *The platform society: Public values in a connective world*. Oxford: Oxford University Press.

Walker, S., Mercea, D., and Bastos, M. 2019. The disinformation landscape and the lockdown of social platforms. *Information, Communication & Society*, **22**(11), 1531–1543.

Walther, J. B. and Jang, J. W. 2012. Communication processes in participatory websites. *Journal of Computer-Mediated Communication*, **18**(1), 2–15.

Youmans, W. L. and York, J. C. 2012. Social media and the activist toolkit: User agreements, corporate interests, and the information infrastructure of modern social movements. *Journal of Communication*, **62**(2), 315–329.

12

Quality and Perspectives

Davide Ceolin, Julia Noordegraaf, and Lora Aroyo

12.1 Introduction

The analysis of perspectives in online information is inherently connected to an assessment of the quality of that information, whereby quality is understood as best meeting the information needs of users. Traditionally, access to information has been regulated by institutional infrastructures that provide clear frameworks for assessing its quality, distinguishing the gossip exchanged on the street by neighbors from the editorially processed information in print and broadcast news media and the highly structured classification of information sources in libraries and archives. The latter infrastructures frame the way we evaluate the relevance of the information for specific needs and also influence the way we assess the perspectives embedded in them.

The emergence of the Internet has complicated this traditional matchmaking between information needs and the relevance of the available documentation through editorial frames. As a result of the participatory nature of the Web, people can contribute different types of information by means of blog posts, articles, tweets, and comments. On the one hand, this has significantly expanded access to information for a broad variety of purposes. On the other hand, on the Web the infrastructures for providing and accessing information are highly diverse and lack a clear framework for the evaluation of the fitness for purpose of the available information and for assessing the perspectives embedded in it.

In particular, the emergence of commercial online platforms such as Facebook, YouTube, and Twitter and the "optimization industry" behind them has influenced the dissemination and evaluation of information, with notorious cases of misinformation and propaganda being circulated through them. The social media mechanisms for sharing and liking introduce "popularity currency" (number of retweets, reposts, likes, etc.), which provides additional motivation for misuse of online platforms to generate artificial popularity of a particular perspective, or to disguise incompetency behind "professional-looking" websites (e.g., low-quality documents can easily be crafted to appear credible and to gain popularity). Behind the scenes, platforms such as Facebook and Twitter employ rules for content moderation that are applied in ways that are nontransparent, with unknown repercussions for the quality and completeness of the information (Gillespie, 2018). In this "disinformation order" (Bennett and Livingston, 2018), where it is hard to distinguish between gossip and information filtered through traditional institutional frameworks, the assessment of the quality of online information becomes a matter of fundamental concern for liberal democracies (Brown, 2018, VanDijck et al., 2018).

In this chapter, we introduce MOIQ (Multidimensional Online Information Quality):[1] a tool for the assessment of the quality of online data. This tool has been developed in the context of the project "Quality and Perspectives in Deep Data" (QuPiD2), which investigates methods and tools for computational support to capture, model, and assess the diversity in quality of online information and the multitude of perspectives (i.e., beliefs, opinions, and world views) reflected therein. Based on quality indicators that have been developed in Library and Information Science – a sector representing public institutions with a longstanding commitment to upholding certain values regarding trustworthiness and reliability of publicly accessible information – the MOIQ tool can support the efforts to strengthen the gatekeeping role of media institutions that should counter the destabilizing effects of the "disinformation order" that Bennett and Livingston (2018) describe. As such, it fits within the wider landscape of tools that support the assessment of the quality of online information, such as the automated fact-checking technologies that assist human fact-checkers in detecting and identifying claims and assessing their veracity (Graves, 2018). Fact-checking tools aim to identify and verify claims in the content itself. MOIQ complements this process by presenting users with metrics on certain quality parameters at the level of the entire document, allowing them to consider whether or not they want to include the source in their analysis.

The assessment of the quality of online information is a difficult task that requires close interaction between human and automated quality assessment workflows and tooling (Graves, 2016, 2018). In addition to the abovementioned challenges of the new information distribution mechanisms, quality is not a monolithic and binomial thing: it is hardly possible to judge documents as "good" or "bad" in absolute terms. The overall quality of a document depends both on the topics, the user that assesses it, and on the intended uses for this

[1] MOIQ is available at `https://moiq.project.cwi.nl/`.

document. However, it is possible to decompose quality into objective "dimensions" or "aspects" that can be combined in order to increase the awareness of possible aspects related to information quality and also to increase the awareness in terms of the relation between quality and perspectives (Son et al., 2016; Fokkens et al., 2016; Ceolin et al., 2016a,b; Van der Zwaan et al., 2016; Ceolin et al., 2016c; Maddalena et al., 2018). Here, we present work on an online information quality assessment tool that incorporates dimensions of the traditional frameworks for the assessment of the quality of documents while taking a human-machine interaction approach in order to suit the scale of the Web.

In general, we can say that **Perspectives** represent the point of view of the author, while **Quality** captures the point of view of the user and, therefore, is subjective and contextual. Information quality is related to the type of information that a given document or item contains and relates this to the reader's requirements. Thus, the two points of view (of the author and of the reader) can be more or less explicitly related. For example, consider a document that uses strong language to claim that vaccination causes autism without citing any sources. Then, consider another document, having a more neutrally phrased text denying that same relation with reference to a variety of peer-reviewed sources and at the website of a governmental medical authority. According to the traditional standards of evaluating information in mainstream media and Library and Information Science, the first document may be evaluated as less accurate and complete and having a stronger bias toward a specific stance than a more neutrally phrased text denying that relation with reference to a variety of peer-reviewed sources, at the website of a governmental medical authority. For a medical journalist of a quality newspaper writing an article on the state of knowledge on the side effects of vaccination, the latter source may be judged of higher quality than the former. On the other hand, for a journalist researching the public debate on vaccination, the perspectives on the topic in both sources may be equally relevant, resulting in high-quality scores for both texts. This example shows that the quality criteria of traditional institutional information evaluation frameworks may be used to inform a range of context-dependent quality assessments.

The chapter continues as follows. Section 12.2 describes the information quality dimensions we focus on. Section 12.3 references related work in this area. Section 12.4 describes a tool we developed and a set of analyses we performed to assess information quality online. Section 12.5 concludes.

12.2 Information Quality Dimensions

Information quality is often defined as fitness for purpose. This informal definition is rather vague, as users can define basically an infinite number of combinations of purposes and fitness functions. In order to operationalize such a definition and to allow to quantify quality in such a way to serve a variety of potential users, diverse quality dimensions have been defined. These quality dimensions narrow the focus on specific aspects of quality and facilitate the definition of metrics to quantify such aspects.

Several quality dimensions have been proposed to measure quality aspects that are relevant in diverse environments. Also, standards have been developed in order to provide a full extension and classification of quality dimensions related to data. For example, the ISO model 25012 for Standardization (International Organization for Standardization, n.d.) describes a set of dimensions for determining data quality. Such definitions focus on low-level and technical aspects of information (e.g., how portable a given piece of data is). In our case, we are interested in higher-level aspects of data and of the information that data represent; we are interested in how users perceive the quality of such data. Thus, we focus on a subset of the quality dimensions described in the ISO model, and we extend and adapt them to cover the main peculiarities of information related to how users perceive and use it. In particular, we identify the following quality dimensions.

Accuracy Indicates the level of perceived truthfulness and veracity of the document presented. This dimension is quantified by crowd workers and expert users, and we use their assessments to train our predictors. However, in the future, we plan to make use of automated fact-checking tools to analyze the accuracy of a document by identifying and verifying the claims made in it.

Precision Indicates the level of precision (as opposed to vagueness) of the language and of the information provided in the document.

Completeness Indicates whether a given document provides a full account with respect to a topic of interest. Again, we capture the level of completeness of the document as experts and crowd workers perceive it.

Relevance Indicates the importance of a given document with respect to a topic of interest. We provide the workers with a description of the topic, and we capture a judgment that quantifies how relevant the document is with respect to the topic according to experts and crowd workers.

Neutrality Indicates the level of bias shown by the document toward a clear stance. In the case of polarizing topics and debates, some documents take clear partisanship, while others take a neutral stance.

Trustworthiness Captures the trust that the reader has in the document source. This is also assessed by experts and crowd workers. The perceived quality of a document is dependent also on the expected trustworthiness in the source that publishes it. This dimension is meant to capture the trust that the experts and crowd workers have in the source itself.

Readability Indicates the level of understandability and comprehensibility of a document according to the human assessors (experts and crowd workers).

Overall Quality Since quality is defined as fitness for purpose, it is not only subjective and contextual but also task-dependent: different tasks imply different purposes. In our studies, we evaluate the fitness of documents with respect to possible tasks at hand. While the other dimensions focus on specific quality aspects, this score provides a unifying view on the quality of the document with respect to its intended use.

It is important to note that since quality assessment is not universally defined, transparency and explainability are key elements of trustworthy quality assessment tools. In fact, there are several factors to consider when collecting multidimensional information quality assessments.

- The list of quality dimensions might be non-exhaustive, and particular tasks might require considering new dimensions. On the other hand, considering some dimensions might be unnecessary for some tasks at hand.
- The assessors might show higher or lower levels of agreements; also, uniform groups of respondents might show a higher or lower reliability depending on their expertise and skills. We will discuss this further in Section 12.4 when comparing crowd- and nichesourced quality assessments.
- The importance of the single dimensions when computing the overall quality score might be diverse. Some dimensions might have a higher importance than others (we will see an example about this in Section 12.4).

All these considerations are important in order to benefit from the subjective nature of quality assessments. If we know who made the quality assessments, and on which premises, then we can judge whether the quality assessments are reliable and trustworthy or not. Note that not all the judgments, items, or dimensions considered are necessarily subjective. Information like the publication date of a document, the identity of the document author, or the result of a fact-checking analysis are likely to be examples of objective information; however, they require a subjective interpretation in order to use them to estimate the quality of the document itself.

12.3 Related Work

The problem of assessing the quality of Web documents, and of (Web) data and information more generally, has been tackled in many contexts.

The ISO 25010 Model for Standardization (International Organization for Standardization, n.d.) is a standard model for data quality. In MOIQ, we make use of the ISO 25010 data quality dimensions that apply also to Web documents (e.g., precision, accuracy). Following up on the use of specific metadata as markers for quality, Amento et al. (2000) use link-based metrics to make quality predictions, showing that these perform as well as content-based ones. In our case, we focus on features we can automatically extract from the documents using AlchemyAPI and WOT.

Regarding the use of niche- or crowdsourcing for collecting information and, in particular, quality assessments, Lee et al. (2002) provide a framework tailored to organizations. Zhu et al. (2011) propose a method for collaboratively assessing the quality of Web documents that shows some similarity with ours (e.g., we both collect collaborative quality assessments). However, the assessments we collect are based on specific tasks.

Digital Humanities scholars are used to critically evaluate the sources they deal with, so we target this specific class of users to investigate how to extend source criticism practices to cover Web documents as well. Source criticism is the process of evaluating traditional information sources that is common in the (Digital) Humanities. De Jong and Schellens (2000) provide an overview of source criticism methods, evaluated in terms of predictive and congruent validity. We will advance such evaluations to identify which Web document features determine their quality.

Lastly, one aspect that we consider when estimating the quality of Web documents is their provenance. In the Humanities, provenance analysis is used to manually assess the quality of sources, as explained by Howell and Prevenier (2001). In Computer Science, Hartig and Zhao (2009) use temporal qualities of provenance traces to assess the quality of Web data. More extensively, Zaveri et al. (2016) provide a review on quality assessment for Linked Data. We also investigated the assessment of crowdsourced annotations using provenance analysis (Ceolin et al., 2016d; Nottamkandath et al., 2015).

12.4 A Tool for Information Quality Assessment

Tool Overview MOIQ is a tool for the assessment of the quality of online documents. It offers the following functionalities: (1) document-centric assessment: for a given URL, MOIQ provides a detailed analysis of its information quality, and (2) topic-centric assessment: for a given topic MOIQ provides an in-depth comparative analysis of the quality of all the documents related to this topic. Our focus is on the assessment of the quality of the information in documents on the Web, such as pages on websites and blog posts. As such, the tool may be used to determine which sources to consult for further analysis at the level of individual statements, for example, as is done by most fact-checking tools. In the future, we intend to expand the tool by testing it on social media content. This requires adapting the tool to detect the quality dimensions in the comparatively much shorter texts of social media posts, which are restricted in length by the platforms on which they are posted. For this task, we may integrate the algorithms of current fact-checking tools that focus on detecting claims in single sentences.

Tool Architecture The quality assessment performed by MOIQ provides a comprehensive, exhaustive, and multi-perspective view along multiple quality dimensions (precision, trustworthiness, accuracy, neutrality, readability, relevance with respect to a given topic). In order to address the intrinsic subjectivity of quality assessment, MOIQ is based on a symbiotic pipeline that brings together humans and machines to gather and train information assessments and the factors that impact them. MOIQ machine learning models are trained on quality assessments provided by crowd contributors and experts. Users were asked to evaluate multiple dimensions (each of them represented on a 1–5 Likert scale) and to motivate their opinions. Opinions have been collected both through ad hoc applications (that allow annotating existing Web documents in their original context) and the dedicated Figure8 platform (which is now called "Appen") (to reach out to crowd contributors).

MOIQ relies on automatic extraction of document features, such as NLP features (e.g., sentiment, named entity recognition) obtained using the Alchemy API (now part of IBM Watson NLP API) and provenance features. In particular, the Web of Trust API (`http://www.mywot.org`) has been employed in order to characterize the trustworthiness of the various sources considered. The API provides trustworthiness scores that are crowdsourced and aggregated in order to consider both the polarity and the size of the samples collected. These features are used to identify correlations between document features and human assessments and allow for the assessment of potentially any Web document. As a machine learning algorithm, we employ multi-label regression and Support Vector Machines. Figure 12.1 gives an overview of MOIQ.

Information Quality Visualization Ultimately, the quality assessments produced by MOIQ are presented to the user by employing a radar chart visualization. On the one hand, this allows the user to obtain a summary of the quality of a document (or of a group of documents) without having to introduce artificial aggregations. Figure 12.2 shows a comparison of five documents of different quality. This overview does not provide details about the exact meaning of each dimension of the radar graph but allows a quick comparison among the assessed

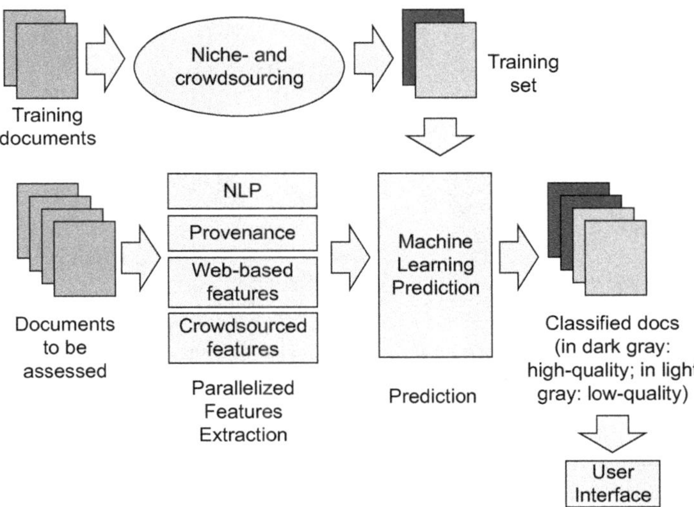

Figure 12.1 Overview of MOIQ, our online quality assessment tool

documents (the smaller the colored area in the graph the lower the quality of the document, and vice versa).

On the other hand, such a visualization allows the final user to investigate further the quality of a given document. When restricting the focus on a lower number of documents, the user can understand precisely how the document scores in each dimension or how two documents compare to each other (see Figure 12.2). This allows increasing user awareness of the quality of documents, in turn, allowing users to decide whether documents meet their contextual and subjective requirements, without, for instance, having to decide which (combination of) quality dimension determines the overall quality of a document.

Tool positioning In the landscape of automated information quality assessment tools, MOIQ positions itself as a tool that aims at providing an overall representation of the quality of documents. It is complementary to automated fact-checking tools, in that it provides an assessment of the broader context in which specific claims are embedded (the document as a whole). Automated fact-checking tools provide useful insights into factual representations and are particularly useful in the case of debates on- and offline. Yet, the identification and verification of claims are complex: for example, some assertions are only partially true or partially false, or evidence on some statements might be scarce. However, when evaluating single statements,

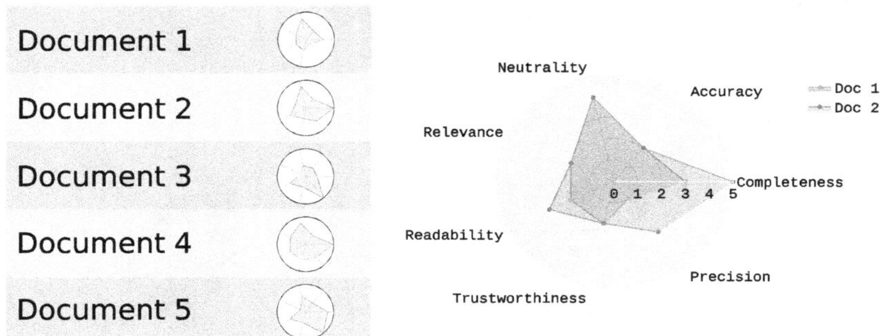

Figure 12.2 Example of a high-level view of a list of assessed documents and of the comparison between two documents

fact-checking them is particularly effective because it provides an intelligible and intuitive indication of the reliability of assertions. When evaluating complex documents, though, the effectiveness of fact-checking is limited. Documents can provide, for instance, only a minor part of false statements, yet the whole meaning of a document might be dictated by these. Also, documents might report only a partial account of correct statements, thus still providing a biased account on a given topic. Therefore, the MOIQ tool provides a complementary perspective on the information quality at the document level.

One of the features that the tool uses to automatically assess the quality of information is the Web of Trust (WOT) score. The WOT score is a crowdsourced score collected by an online platform indicating the maliciousness of Web sources. While this score does not directly indicate the quality of information, it provides a first filter for low-quality documents. In the future, we plan to also incorporate additional information quality signals, including those from fact-checking platforms.

MOIQ has been tested on a group of fifty documents regarding the vaccination debate (selected in order to represent a small but heterogeneous sample consisting of blog posts, news articles, documents from public authorities, etc.), where it shows a promising performance (up to 90% accuracy) that reflects previous works of ours (Ceolin et al., 2016a,c). We envision three main future developments for MOIQ. First, the extension of the domains and topics covered (currently, the tool allows any document to be assessed, but the accuracy of such assessments is under evaluation). Second, the improvement of the computation speed. Third, the personalization of the quality assessments, such that different users can obtain assessments matching their specific needs and requirements.

Tool Evaluation In order to evaluate the tool performance, we ran two user studies involving forty experts in total, half of which were journalism students and the other half media research scholars. We asked these experts to evaluate fifty documents regarding the measles outbreak that happened at Disneyland, California, in 2015. The small corpus of documents was collected in order to give a full account of the vaccination debate sparked by that event. Thus, the corpus contains both pro- and anti-vaccination documents. Also, the documents therein reported are diverse in terms of authoritativeness: the corpus contains news articles, official authority reports, blog posts, and so on. We used MOIQ to collect multidimensional judgments about those documents along the aforementioned seven quality dimensions. Also, we asked the respondents to judge whether the documents met their quality standards in order to be used as a source to write an article about the vaccination debate. This is the task on which the quality judgments are based.

As a result of those studies, we collected 151 assessments of the documents. Thanks to the uniform background of the respondents and to the specific task at hand, the judgments collected are quite uniform. In fact, they show a high correlation among each other, and it is possible to estimate them by extracting textual features from the documents and using the Support Vector Classifier algorithm, reaching an accuracy of 89 percent. Also, the quality judgments are quite dependent on the task at hand: since the respondents were interested in giving a full account of the vaccination debate, the neutrality of the documents was not considered as an important aspect. A full overview of the experimental setting and of the results obtained is available in Ceolin et al. (2016a).

Another conclusion we drew from these two studies is that while they provide insightful assessments of the documents, the tasks are rather demanding and time-consuming. The insights obtained are deep, but they are limited in the number of documents they cover. One solution to this limit is the use of crowdsourcing to increase the workforce at our disposal. We ran a pilot study to test the possibility of using crowdsourcing as an alternative to nichesourcing (Maddalena et al., 2018). Results show that while crowdsourced assessments are sufficiently correlated with each other, they show a low correlation with expert-provided judgments. This is likely to be due to the heterogeneity of the population of crowd workers, as compared to the experts. Therefore, in order to fully embrace the potential offered by crowdsourcing, we will need to identify strategies to discern among respondents. The cognitive reflection tests (CRTs) (Frederick, 2005) represent a promising solution to this problem, as shown in the work of Roitero et al. (2020). This line of research will be pursued further in the future because experts alone can hardly cope with the assessment of the vast multitude of documents produced online. This problem is also exacerbated by the fact that the assessments may have a limited temporal validity: new discoveries, insights, debates may lead to revisions on the truthfulness of previously assessed statements and documents. Thus, this requires a very large human workforce, and properly instructed and selected crowd workers show a promising solution to this issue.

12.5 Conclusion and Future Work

In this chapter, we presented and discussed MOIQ – a tool assessing the quality of Web documents from a multidimensional standpoint. The tool is based on a combination of nichesourcing (i.e., of expert-based crowdsourcing), crowdsourcing, and machine learning and provides assessments on the quality of online documents. It has been tested on a sample of fifty online documents on the vaccination debate. In particular, MOIQ assesses the quality of documents considering their accuracy, precision, completeness, relevance, neutrality, and trustworthiness. Also, given that quality is contextual and task-dependent, in our analyses, we introduce a hypothetical task for the experts involved. Experts were provided with a selection of documents to evaluate and, after having assessed them, were asked to determine whether they would consider each document to be used as a source for an article on the vaccination debate. The resulting decision can be seen as an overall quality assessment, unifying the multidimensional assessments into a single score. Our analyses show that the expert-based assessments are regular enough to be automatically predicted with promising accuracy.

As future development, we expect to scale up both computation streams: human computation and machine learning. Regarding human computation, we are exploring the use of crowdsourcing as a means to increase computation availability and compensate for the limited availability of experts. Crowdsourcing requires additional quality checks, as crowd workers are less uniform than experts, and thus, the quality of their assessments is diverse. However, the incorporation of crowdsourcing in our pipeline will be further investigated to cope both with the high volume of Web documents and with the changing nature of truthful assessments. Moreover, we will explore the use of natural language processing tools in combination with linked data to provide rule-based systems that both provide accurate quality assessments and accompany them with intelligible explanations and motivations. Lastly, MOIQ will be extended into two additional directions. First, the system will become capable of assessing social media items. This requires tackling shorter pieces of text and leveraging information coming from the surrounding social media environment (interactions, network positioning, etc.). Second, fact-checking tools and, in particular, their output will be integrated into our system too. These tools are a particularly valuable source of trusted information that is provided at a granular level. Checks and verification about statements and claims then need to be properly handled in order to provide insights into the overall quality of documents that are inherently more complex in nature.

References

Amento, B., Terveen, L., and Hill, W. 2000. Does authority mean quality? Predicting expert quality ratings of Web documents. *SIGIR, ACM*, 296–303.

Bennett, W. L. and Livingston, S. 2018. The disinformation order: Disruptive communication and the decline of democratic institutions. *European Journal of Communication*, **33**(2), 122–139.

Brown, É. 2018. Propaganda, misinformation, and the epistemic value of democracy. *Critical Review*, **30**(3–4), 194–218.

Ceolin, D., Noordegraaf, J., and Aroyo, L. 2016a. Capturing the ineffable: Collecting, analysing, and automating Web document quality assessment. *EKAW*, 83–97.

Ceolin, D., Aroyo, L., and Noordegraaf, J. 2016b. Identifying and classifying uncertainty layers in Web document quality assessment. *URSW*, 61–64.

Ceolin, D., Noordegraaf, J., Aroyo, L., and van Son, C. 2016c. Towards Web documents quality assessment for digital humanities scholars. *In WebSci*, **16**, 315–317.

Ceolin, D. Groth, P., Maccatrozzo, V., Fokkink, W., van Hage, W.R., and Nottamkandath, A. 2016d. Combining user reputation and provenance analysis for trust assessment. Journal of Data and Information Quality, *ACM*, **7**(1–2), 6–1.

De Jong, M., and Schellens, P. J. 2000. Toward a document evaluation methodology: What does research tell us about the validity and reliability of evaluation methods? *IEEE Transactions on Professional Communication*, **43**(3), 242–260.

Fokkens, A., Ter Braake, S., Maks, I., and Ceolin, D. 2016. *On the semantics of concept drift: Towards formal definitions of semantic change*. In Proceedings of the Drift-a-LOD Workshop. Balogna: CEUR.

Frederick, S. 2005. Cognitive reflection and decision making. *Journal of Economic Perspectives*, **19**(4), 2005.

Gillespie, T. 2018. *Platforms, content moderation, and the hidden decisions that shape social media*. New Haven, CT: Yale University Press.

Graves, L. 2016. *Deciding what's true: The rise of political fact-checking in American journalism*. New York: Columbia University Press.

Graves, L. 2018. Understanding the promise and limits of automated fact-checking. Reuters Institute for the Study of Journalism, University of Oxford: Fact Sheet.

Hartig, O. and Zhao, J. 2009. *Using Web data provenance for quality assessment*. In Proceedings of the 8th International Semantic Web Conference, CEUR Workshop. Washington: SWPM.

Howell, M. and Prevenier, W. 2001. *From reliable sources: An introduction to historical methods*. Ithaca, NY: Cornell University Press.

International Organization for Standardization. n.d. ISO model 25012 for Standardization. ISO/IEC 25012:2008: Software engineering – Software product Quality Requirements and Evaluation (SQuaRE) – Data quality model. https://www.iso.org/standard/35736.html.

Lee, Y. W., Strong, D. M., Kahn, B. K., and Wang, R. Y. 2002. AIMQ: A methodology for information quality assessment. *Information Management*, **40**(2), 133–146.

Maddalena, E., Ceolin, D., and Mizzaro, S. 2018. *Multidimensional news quality: A comparison of crowdsourcing and nichesourcing.* In Proceedings of the 6th International Workshop on News Recommendation and Analytics 4. New York, NY: ACM.

Nottamkandath, A., Oosterman, J., Ceolin, D., de Vries, G. K. D., and Fokkink, W. 2015. *Predicting quality of crowdsourced annotations using graph kernels.* In Proceedings of the International Conference on Trust Management, 134-148. New York, NY: Springer.

Roitero, K., Soprano, M., Fan, S., Spina, D., Mizzaro, S., and Demartini, G. 2020. *Can the crowd identify misinformation objectively? The effects of judgment scale and assessor's background.* In Proceedings of the 43rd International ACM SIGIR Conference on Research and Development in Information Retrieval (SIGIR '20). New York, NY: Association for Computing Machinery, pp. 439–448.

Son, C. van, van Miltenburg, E., and Morante, R. 2016. *Building a dictionary of affixal negations.* In ExProM 2016.

Van der Zwaan, J. M., van Meersbergen, M., Fokkens, A., ter Braake S., Leemans I., Kuijpers E. Vossen P., and I., Maks. 2016. *Storyteller: Visualizing perspectives in digital humanities projects.* 2nd IFIP Workshop on Computational History and Data-Driven Humanities.

Van Dijck, J., Poell, T., and Waal, M. de. 2018. *The Platform Society: Public Values in a Connective World.* Oxford: Oxford University Press.

Zaveri, A., Rula, A., Maurino, A., Pietrobon, R., Lehmann, J., Auer S. 2016. Quality assessment for linked data: A survey. *Semantic Web*, **7**(1), 63–93.

Zhu, H., Ma, Y., and Su, G. 2011. Collaboratively assessing information quality on the Web. *ICIS sigIQ Workshop.*

PART IV

Mining and Modeling Perspectives

Section Editors: Piek Vossen and Antske Fokkens

13

Mining and Modeling Perspectives

Piek Vossen and Antske Fokkens

To monitor, analyze, and understand the content and the dynamics of the abundant communication on the Web, we need natural language processing (NLP). Various techniques have been developed in the last decades for different aspects of communication, such as sentiment and emotion detection, opinion mining, attribution detection, argument mining, and more recently, hate speech and fake news detection. Most of these techniques use annotated corpora and/or lexicons for different languages, and often machine learning approaches give state-of-the-art results. Different genres of data have been annotated with interpretation labels by trained experts or the crowd and various machine learning methods used for creating classifiers with high scores (up to 90% accuracy and more) on separated test data, usually similar to the training data.

However, perspectives typically form a more complex relation between the "posters" (sources) of texts and the message sent out to a presumed audience. Especially the nature of social media, in which many people post independently both on global platform and within specific bubbles that accelerate opinions, makes it difficult to connect messages and to take the context into account. We therefore need to combine various techniques and we need more complex models than generating labels for texts, for example, positive or negative tweet classification. Communication is a complex social and cognitive phenomenon. We use language to ventilate complex perspectives within specific social contexts and it is not easy for machines to interpret texts without such contexts. For example, sarcasm, irony, and toxic language are abundant on a medium such as Twitter, as well as trolling and misinformation. It is, however, not just a matter of determining who tweets what but merely why does who tweet when. Such implicit data is difficult to recover and detect.

Part IV of this book discusses NLP and Semantic Web technologies to grasp the complex phenomena of perspectives and their application to service users of the Web in its full complexity. The six chapters operationalize the concept of perspectives in terms of relations and models that can be detected and reasoned upon computationally. Various specialized components are discussed that can analyze texts to reveal debate structures. Specific attention is paid to the integration of these analyses into models that capture perspectives of people with respect to their claims. Finally, the chapters in Part IV address various data sets on specific debates, illustrating the problem and solutions through concrete text examples and structures but also providing evaluations on annotated data that show how advanced the technology is.

Chapters 14 and 15 have a strong focus on the NLP techniques needed for the detection and modeling of perspectives. Chapter 14 gives an overview of the state-of-the-art in NLP on three key phenomena: attribution, factuality, and propositional alignment. NLP approaches usually address these as separate tasks, but in Chapter 19, these are combined in one model. This model is applied to a corpus with texts on the vaccination debate that contains diverse and dispersed claims on a single conceivable topic. Chapter 15 describes the use of NLP techniques to aggregate a broad range of perspectives for users in search of context for claims. Here the focus is on discovering perspectives from online sources and providing a comprehensive overview as a proxy for what is on the Web as opposed to biased results due to personalization or popularity. Both chapters discuss what NLP is needed to mine perspectives and gather supporting evidence to give people a more complete and transparent insight into a debate. Chapter 16 tries to model the output of NLP systems formally to enable reasoning of perspectives. It uses an explicit resource of conceptual frames to model perspectives.

Chapters 17 and 18 reflect on applications that use perspective models to provide transparency. Both chapters address biases in applications and data but in different ways. Chapter 17 discusses perspectives in the context of information retrieval for professional purposes, where users want to be in control of the search process without suffering from unwanted biases. Chapter 18 looks into actual topics of hate speech and toxic language detection, especially on social media, which results in various biases of data and systems derived from this data. The authors propose a distant-supervision approach to circumvent such biases by getting more balanced perspectives.

Part IV of this book gives good insight into the depth and complexity of the phenomena from a more technical point of view. Although the problem is traditionally broken down into smaller subproblems and components, the approaches described here put these together and try to give a complete picture of where technology stands in relation to the human online debate.

14
Natural Language Processing Tasks for the Extraction of Perspectives

Chantal van Son, Roser Morante, and Piek Vossen

14.1 Introduction

The Internet acts as a global forum where everybody can contribute to any topic via different modalities, one of which is textual. On a daily basis, millions of authors share their knowledge, thoughts, ideas, experiences, desires, stories, opinions, and views on the world through text. Therefore, these texts not only provide a rich resource of information, but are also a reflection of ongoing debates in our society, stances on particular issues (e.g., abortion, vaccinations, etc.), and interpretative frames on events and their causes (e.g., conspiracy theories on COVID-19). In addition, while many of the statements are reliable and well motivated, there are also a lot that lack evidence, are based on misquotations, or simply present fake facts. Tools that can automatically process, categorize, and analyze the complex mixture of knowledge and opinions could prove extremely valuable in supporting Web users in navigating through this overwhelming information landscape.

Natural language processing (NLP) can contribute to building a "web of perspectives" by automatically processing textual information and extracting perspectives (Chen et al., 2019; Vilares and He, 2017). However, the concept of perspective, the means to model perspectives, and the procedures to extract them are yet to be determined. Some research has been published, but the extraction of perspectives is not yet a consolidated task. In this chapter, we aim to contribute to this goal by presenting our view on perspectives and some of the existing NLP tasks that can play a role in extracting them.

An initial model of perspectives needs to represent at least the following elements: (1) statements or claims, (2) the sources of statements, and (3) any indication of subjective evaluation of that statement, such as negation, (un)certainty, emotion, or sentiment. Hence, we define a perspective as the attitude of a source toward a statement expressed by means of multiple perspective values. In addition to representing this relation between source and statement, a complete model of perspectives would be able to contextualize them by comparing the perspectives with other perspectives on the same or related statements. This requires that we also model the *content* of statements, enabling true understanding of the perspectives and the relations between them. In Chapter 19, a formal model is given to represent these relations following the Grounded Representation and Source Perspective framework. This chapter focuses on some of the NLP techniques that can be used to extract perspective relations from texts.

A generic and comprehensive perspective mining system should integrate the various existing NLP tasks that focus on the detection of events, attribution, negation, hedging, (un)certainty, factuality, emotion, opinion, argument structure, stance, and other phenomena related to subjectivity in text. Our ongoing work involves annotating these different layers of perspectives in the Vaccination Corpus (Morante et al., 2020),[1] a corpus of texts (news, blogs, editorial, governmental reports, science articles) related to the online vaccination debate that has been compiled to study the language of online debates. So far, we have annotated vaccine-related events, attribution, claims, and opinions. In this chapter, we limit our discussion to a specific subset of NLP tasks that contribute to detecting levels of truth and factuality as perceived by different people and sources. Which claims are made with absolute certainty, and which are presented with some level of cautiousness with respect to their truth? What do people agree upon? Which claims are questioned?

Three tasks that contribute to answering these questions are attribution processing, factuality detection, and natural language inference (NLI). Attribution processing focuses on linking the source of a propositional attitude to its target statement, factuality detection further characterizes this propositional attitude by specifying the level of commitment of the source toward the statement, and NLI aims at identifying entailment or contradiction relations between statements. In addition to providing a summary of how these three tasks have been tackled in the computational linguistics community, illustrating them with examples from the Vaccination Corpus, we aim to stress two aspects that we feel are crucial yet sometimes overlooked.

First, we underline the importance of attribution for both factuality detection *and* NLI as downstream tasks. To illustrate, do Sentences 1 and 2 below contradict or agree with each other? A human reader will probably recognize two different attitudes expressed toward the proposition *vaccines cause autism* in

[1] The annotations of the Vaccination Corpus are available at `https://github.com/cltl/VaccinationCorpus`.

Sentence 2, namely that of the author and that of *some parents*, and therefore that the author of Sentence 2 would agree with Sentence 1, but *some parents* would not.

1. Vaccines are very safe.
2. *Some parents* still believe [the debunked theory that vaccines cause autism].

While the phenomenon of additional sources introduced in the text has been acknowledged in factuality research, most corpora focus only on the author's perspective. In the NLI task, the presence of quoted sources has largely been ignored. Historically, NLI has mostly been performed on artificially created data to simplify an extremely complex task. However, if we were to move to testing our systems on more realistic data, we will have to deal with attribution, especially in the context of online debate.

Second, we argue that many statements contain smaller propositional units whose factual status can be judged independently from the full proposition. We call these **micro-propositions**. Acknowledging the notion of micro-propositions has two implications. First, linguistic means to express negation or uncertainty may target only part of a proposition, affecting tasks like factuality detection. For instance, Sentences 3, 4 and 5 all discuss the effectiveness of vaccines, but not in absolute terms. The targets of the attitudes are specific aspects or scenarios regarding their effectiveness, that is, for whom they are effective (*patients with HIV*), when they are effective (*after childhood*), and the extent to which they are effective (*100%*). Second, the independent status of these micro-propositions means that each of them is up for discussion. Other sentences may therefore agree with the factual status of some, but not all, which is relevant for tasks like NLI.

3. Hepatitis B vaccination is less effective in patients with HIV.
4. The vaccination is not quite as effective after childhood.
5. No immunisation is 100% effective.

Therefore, we think that it is important to integrate micro-propositions in models of factuality and natural language inference. Micro-propositions allow us not only to capture more specific interpretations of perspectives but also to compare different perspectives toward statements in and across texts in a flexible way. For this purpose, we treat a sentence as a *bag-of-micro-propositions* for which the precise internal relation may not be explicit but different *bags* can be compared for their components.

The rest of the chapter is structured as follows. First, we provide an overview of how existing annotation models capture different aspects of attribution (Section 14.2), factuality (Section 14.3), and NLI (Section 14.4), and what are some of the main approaches to process them automatically. Section 14.5 introduces the concept of micro-propositions in more detail and discusses how this phenomenon has previously been modeled in factuality detection (Section 14.5.1) and negation processing (Section 14.5.2). We conclude in Section 14.6.

14.2 Modeling and Processing Attribution

An essential task when processing perspectives is to determine who is the holder of the perspective and what the perspective is about. There is an NLP task that focuses on finding what is called **attribution relations** (ARs). Generally speaking, an attribution happens when someone attributes a statement, a belief, a feeling, an intention, or any other attitude to a source. Technically speaking, an AR is established when someone (the author of a text, someone mentioned in the text) signals an ownership relation between a source (a third party who expresses an attitude) and some text. ARs have been defined as relations that ascribe the ownership of an attitude toward some linguistic material, that is, the text itself, a portion of it or its semantic content, to an entity (Pareti, 2015; Prasad et al., 2007). Thus, **attribution processing** is an NLP task that consists in finding all ARs in a text and labeling the spans of text that constitute the elements of the AR.

An AR is composed of three elements (Pareti, 2015; Prabhakaran et al., 2010; Prasad et al., 2007):

- A *source*: the owner of the attitude. It can be realized by named entities (persons, organizations, countries), proper nouns, or noun phrases.
- A linguistic *cue* that explicitly introduces the AR.
- A *content*: the span of text attributed to the source. This span of text expresses an attitude.

Sentences 6–7 provide examples of attribution relations:[2]

6. And *we* **know** [that vaccinations will save many, many lives].
7. ["Although mercury has been removed from many vaccines, other culprits may link vaccines to autism,"] **said** *the study's lead author*.
8. *People* **blame** [the park for the measles outbreak].

ARs can be nested and the content span can be discontinuous, as shown in (9) and (10), respectively.

9. [While *Trump's team* **denied** [making the specific request that Kennedy referenced], *a spokeswoman* **said** [that Trump is considering forming a commission on autism]], **reported** *CNN and other outlets*.

[2] The source is marked in italics, the cue in bold and the content is between square brackets.

10. [In order for full protection to be gained,] **claims** *the establishment*, [a 95% vaccination rate is required for vaccines that are 100% effective].

ARs may have three different surface realizations, as illustrated in the examples below: direct reported speech (signaled by quotation marks) (11); indirect reported speech (12); and mixed reported speech (13).

11. *They* **concluded**: ["...measles outbreaks can occur among highly vaccinated college populations."]
12. *The news story* **emphasized** [that whooping cough is highly dangerous and can lead to vomiting and death, especially in children].
13. In late January, *the Centers for Disease Control and Prevention's Dr. Anne Schuchat* **said** [the uptick in measles cases "is a wake-up call to make sure measles doesn't get a foothold back in our country."]

In order to train and evaluate systems that can perform the attribution labeling task automatically, annotated corpora are needed. Two publicly available corpora exist for English, the Penn Attribution Relations Corpus (PARC) (Pareti 2012) and PolNeAR (Newell et al., 2018)[3]. PARC is a collection of 2,280 *Wall Street Journal* articles, whereas PolNeAR contains 1,008 news articles that cover the presidential candidates Hillary Clinton and Donald Trump during the campaign of the 2016 US presidential elections.

Computational work on processing ARs focuses on extracting direct quotations (Almeida et al., 2014; de La Clergerie et al., 2011; Elson and McKeown, 2010; Muzny et al., 2017; O'Keefe et al., 2012; Schneider et al., 2010), indirect quotations (Weiser and Watrin, 2012), or both (Pareti et al., 2013; Scheible et al., 2016). Extracting direct quotations is easier due to the presence of quotation marks that signal the attribution. Most work has been done for English.

A typical system for direct quotation extraction identifies direct quotations and speakers and assigns a speaker to each quotation (Almeida et al., 2014; Elson and McKeown, 2010; O'Keefe et al., 2012). In some cases, coreference resolution is integrated into the system (Almeida et al., 2014). The standard system to extract and label attributions uses three machine learning classifiers, one for each component (source, cue, and content) (Pareti et al., 2013). More recent work reports on different architectures. Scheible et al. (2016) explore two architectures for quotation detection, one where token-level classifiers predict quotation boundaries and combine the boundaries greedily to predict spans, and another one where a semi-Markov sequence model uses global features of

[3] Website of the PolNeAR corpus: `https://github.com/networkdynamics/PolNeAR`.

quotation spans.[4] Finally, Muzny et al. (2017) present a tool for quotation annotation allowing for the annotation of quotes in literary texts and the speakers that they are linked to.[5]

14.3 Factuality Profiling

Factuality Profiling is an NLP task that consists of determining the degree of commitment that sources of statements hold in relation to the propositions contained in these statements. Broadly speaking, while attribution processing focuses on linking a linguistic expression to its source (*who* says or believes the statement), factuality profiling further characterizes this relation by determining the epistemic perspective of the source in relation to their statements. Do they know whether something happened or not? Are they sure about what they are saying? Are they talking about events that actually happened or about hypothetical events? For example, consider the sentences below. While they all contain the same proposition, `cause(vaccines, autism)`, the author expresses different levels of commitment toward it. The proposition may be confirmed (14) or denied (17) by the author, and the author may express different levels of certainty (15 and 16).

14. Vaccines cause autism.
15. Vaccines **may** cause autism.
16. Vaccines **may not** cause autism.
17. Vaccines **don't** cause autism.

A definition of factuality was first given by Saurí (2008), who described it as "the level of information expressing the commitment of relevant sources toward the factual nature of eventualities in text." Several corpora have been annotated with factuality-related information to enable the training and evaluation of automated NLP systems to detect event factuality.

There are a couple of aspects to the notion of factuality that have been recognized as crucial for modeling and annotating the phenomenon in natural language texts. The first aspect concerns the different dimensions that are involved in characterizing the factual status of an event. A variety of linguistic means are available to language users to communicate their doubts, certainties, and guesses. For instance, negation (e.g., *not*, *never*, *nobody*) is used to

[4] Tool available at
`www.ims.uni-stuttgart.de/forschung/ressourcen/werkzeuge/qsample.html`.

[5] Tool available at
`https://stanfordnlp.github.io/CoreNLP/quoteattribution.html`.

indicate that something is *not* the case, epistemic (e.g., *could, might*) and deontic (e.g., *should, need*) modality markers express some degree of probability or necessity associated with the proposition, and evidentiality markers (e.g., *I saw that…, I was told that…*) indicate the evidence a speaker has for their statement. There is no single answer to the question as to how these phenomena should be conceptualized into a discrete set of values that can be used for a standardized classification task. As a result, different annotation schemes have been proposed to capture this level of information.

FactBank (Saurí and Pustejovsky, 2009), for instance, models factuality as the combination of *polarity* and *certainty* and uses a set of six core values for its annotation:

CT+	certainly the case	**CT–**	certainly not the case
PR+	probably the case	**PR–**	probably not the case
PS+	possibly the case	**PS–**	possibly not the case

In addition, FactBank includes two values **CTu** and **Uu** to respectively account for cases where the source is certain about the factual nature of the event, but it is not clear what the output is (e.g., *John knows whether Mary came*), and cases where the source does not know what the factual status of the event is or does not overtly commit to it (e.g., *John does not know whether Mary came*). A similar set of values was used for annotating the MEANTIME corpus (Minard et al., 2016), where each event was associated with a factuality value described through several attributes, including polarity, certainty, and time. The temporal aspect was added to distinguish between past/present events and future events, where the actual status of the latter is by definition uncertain (van Son et al., 2014).

The Committed Belief framework used to annotate the Language Understanding (LU) corpora (Diab et al., 2009; Prabhakaran et al., 2010; Werner et al., 2015) takes a less granular approach and uses only four tags to annotate their corpus:[6]

Committed Belief (CB)	the author strongly believes it is true
Non-Committed Belief (NCB)	the author weakly believes it is true
Non-Attributable Belief (NA)	another type of attitude
Reported Belief (ROB)	another source than the author

[6] The first version of this annotation scheme only had three labels: CB, NCB, and NA. However, Werner et al. (2015) argued that the NCB category in fact captured two different notions: that of uncertainty of the author and that of belief being attributed to someone other than the author, and therefore manually relabeled this category with the additional ROB tag.

Instead of using some fixed categories, the UW dataset (Lee et al., 2015) was built by asking crowd workers to score each event on a scale of −3 (*certainly did not happen*) to 3 (*certainly did*), and aggregating multiple judgments to get the final label. This UW scale of [−3, 3] was adopted by Stanovsky et al. (2017), who mapped the discrete annotations of both FactBank and MEANTIME onto this scale to supplement the UW dataset, resulting in the unified factuality (UF) corpus. Finally, factuality annotation in the Universal Decompositional Semantics It Happened datasets, UDS-IH1 (White et al., 2016) and UDS-IH2 (Rudinger et al., 2018) were collected through crowd annotation by asking workers who identified a particular word as an event (or state) whether or not, according to the author, the event has already happened or is currently happening, and how confident the annotator is about their answer on a scale from 0 to 4. Important to note here is that the latter question concerns annotator confidence rather than source confidence, which is different from the previous annotation schemes.

Another aspect that should be taken into account when modeling factuality is that it should always be assessed relative to a specific **source**. Naturally, when reading a text or listening to someone speaking, we interpret the belief of the author or speaker toward the status of the statement made. However, the author or speaker may also introduce additional sources into the discourse that are relevant for assessing the factuality of an event by means of what Saurí and Pustejovsky (2009) refer to as source-introducing predicates (SIPs), such as predicates of reporting (e.g., *say*, *tell*), knowledge and opinion (e.g., *believe*, *know*), psychological reaction (e.g., *regret*), and so on. For instance, the verb *believe* in Sentence 18 is used to attribute the belief regarding vaccines not producing any side effects to *doctors and nurses*.

18. There are *doctors and nurses* around who **believe** [that vaccines do not produce any side effects at all].

The author or speaker may explicitly agree or disagree with the factuality evaluation of their quoted sources, or they may remain uncommitted themselves and leave the final interpretation up to the reader (as is often the case in news reports, for instance). There may even be multiple layers of nesting (*X said that Y believes...*), distancing the author even further from any judgment. Factuality profiling thus requires making explicit *to whom* we attribute a certain belief since assuming that the same evaluation applies to the different sources may be conceptually insufficient or even incorrect. Therefore, there is a strong connection between this line of research and processing attribution.

Different approaches have been taken to deal with the multiple perspectives that may be expressed toward a single event, where most corpora focus only on the author's point of view (Diab et al., 2009; Lee et al., 2015; Prabhakaran et al., 2015; Rudinger et al., 2018; Werner et al., 2015; White et al., 2016) and only some annotate a single event with multiple assessments for each of the relevant sources (Ellis et al., 2016; Saurí and Pustejovsky, 2009). When events are assessed for multiple sources, these are typically encoded as (chains of) nested sources to acknowledge the fact that, strictly speaking, we cannot be certain of the *actual* commitment of quoted sources (all we know is what the author tells us). For instance, the event *produce* in Sentence 18 would receive two values in FactBank, Uu for *author* and CT– for *doctors-and-nurses_author*, where the latter should be interpreted as: "according to the author, *doctors and nurses* believe that it is `certainly not the case` that *vaccines produce any side effects*." Finally, differences in factuality judgements may also arise on the interpretation side of communication; De Marneffe et al. (2012) captured this pragmatic aspect of factuality by supplementing a subset of FactBank with the *reader's veridicality judgment* of events using crowd annotation, arguing that the full distributions of judgments are actually more appropriate for modeling veridicality judgments than a single expert-annotated label.

Methods to predict factuality in an automatic manner can be largely divided into rule-based systems which examine deep linguistic features, and machine learning algorithms which generally extract more shallow features. Both the De Facto factuality profiler (Saurí, 2008) and the TruthTeller algorithm (Lotan et al., 2013) assign discrete factuality values to events and predicate using a rule-based approach on dependency trees that relies on a handwritten lexicon of predicates and other lexical cues (negation, modality), indicating how the factuality assessment of their embedded predicates should be modified when they are encountered. Diab et al. (2009) and Prabhakaran et al. (2010) train support vector machine (SVM) and conditional random fields (CRF) models with lexical and syntactic features (lemma, part-of-speech, dependency paths) to solve the classification problem, while Lee et al. (2015) follow a similar approach using SVM regression techniques to predict the continuous factuality value in the UW corpus. Stanovsky et al. (2017) use a combination of rule-based and supervised approaches by training an SVM regression model on top of the output of an extended version of the TruthTeller lexicon and algorithm. Finally, Rudinger et al. (2018) tackle the regression task by developing two neural models of event factuality prediction: a bidirectional linear-chain LSTM (L-biLSTM) and a bidirectional child-sum dependency tree LSTM (T-biLSTM).

14.4 Natural Language Inference

The **natural language inference (NLI)** task aims to characterize relations between sentences by determining whether the truth of one sentence (the hypothesis *h*) entails from, contradicts with, or is logically neutral with respect to another sentence (the premise *p*). For instance, Sentence 19a entails Sentence 19b, Sentence 20a contradicts Sentence 20b, and there is a neutral relationship between Sentences 21a and 21b.

19. a. Vaccines cause autism.
    b. Vaccines are dangerous. ENTAILMENT
20. a. Vaccines cause autism.
    b. Vaccines are safe. CONTRADICTION
21. a. Vaccines cause autism.
    b. Vaccines are expensive. NEUTRAL

Early work on NLI was mostly driven by the recognizing textual entailment (RTE) evaluation challenges (Dagan et al., 2006).[7] Depending on the application setting (e.g., Information Retrieval, Question Answering, Reading Comprehension), there was some variety in the ways in which the premise-hypothesis pairs in these RTE datasets were collected, but in most cases, the hypotheses were constructed by the annotators and complemented with premises by carefully examining different text snippets. This resulted in high-quality, hand-labeled datasets that have greatly stimulated the development of innovative inference models.

In recent years, however, researchers started to look for alternative methods to generate premise-hypothesis pairs in a more effective or more controlled way. This has resulted in a plethora of new and often larger NLI datasets (Bowman et al., 2015; Khot et al., 2018; Lai et al., 2017; Marelli et al., 2014; Nangia et al., 2017; Poliak et al., 2018a). What connects many of these corpora is that the premises are usually selected from some existing corpus (e.g., image and video descriptions, Web texts) and taken as they are, while the hypotheses are either elicited from the crowd or automatically generated by manipulating and simplifying sentences through lexical and syntactic transformations.

While these construction methodologies have made it possible to create NLI corpora large enough to train deep neural network models, recent studies have shown that they have also resulted in data biases that make it possible for a model without access to the premise to correctly predict the label

[7] `https://aclweb.org/aclwiki/Textual_Entailment_Resource_Pool`.

on the basis of the hypothesis only (Gururangan et al., 2018; Poliak et al., 2018b; Tsuchiya, 2018). For instance, specific annotation strategies and heuristics adopted by crowd workers to generate the hypotheses in SNLI (Bowman et al., 2015) and MultiNLI (Nangia et al., 2017) are likely the cause of high correlations between specific linguistic phenomena (e.g., negation, vagueness, word choice) and certain inference classes. In the words of Gururangan et al. (2018), this raises the question of whether state-of-the-art NLI models performing well on these datasets have successfully "understood" natural language or whether they have "gamed" the task by simply exploiting these statistical irregularities.

A possible explanation of why so far NLI research has mostly refrained from characterizing the relations between pairs of *naturally occurring* sentences is that these relations are often highly complex and cannot be reduced into a single label. Consider the following examples:

22. Vaccines are very safe and effective.
23. Don't just take your pediatrician's word that shots are safe.
24. Vaccines are not safe for people with immune problems.

Annotating the sentence pairs 22–23, 23–24, and 22–24 with either NEUTRAL, ENTAILMENT or CONTRADICTION would not do justice to the subtleties in meaning that humans are capable of recognizing. First of all, while Sentence 22 makes a claim on both the effectiveness and the safety of vaccines, Sentences 23 and 24 only discuss their safety and can thus not be said to agree or disagree on their effectiveness. Second, Sentence 23 presents two different views with respect to the safety of vaccines: that of *your pediatrician*, who claims they are safe, and that of the author, who claims they are not. Finally, the most likely interpretation of Sentence 24 is not that vaccines are not safe at all, but rather that vaccines are safe, but not for people with immune problems. This means that this sentence may only partially (dis)agree with the others.

In other words, in order to create more realistic benchmarks that expose the true performance levels of NLI models on natural data, the task will have to be defined more specifically. Before we can even begin to understand and analyze the logical relations between two sentences, we need to decompose sentences into statements and perspectives on those statements. This involves specifying exactly (a) which propositions within the sentence we are comparing, (b) whose perspectives (the authors' and/or quoted sources') we are comparing, and (c) whether the perspectives on those statements completely or only partially (dis)agree. This is where the role of attribution and factuality detection becomes evident as an essential first step toward solving the NLI task.

14.5 Micro-Propositions

In this section, we introduce the concept of micro-propositions and explain their relevance in the context of factuality profiling and NLI. We define **micro-propositions** as the smaller meaningful propositional units which factual status can be judged independently from the full proposition they are part of. This independent status of micro-propositions means that each of them is also up for discussion. For example, consider Sentence 25 below. We can imagine how the truth of this claim may be contested in a variety of ways. For instance, someone might counter this claim by saying that the smallpox inoculation was banned in 1841 instead of 1840, or that it was some other type of inoculation that was banned in England that year. Similarly, another statement may only confirm the location where this inoculation was banned, but not the year. In other words, it is often the case that agreement or disagreement only arises on specific aspects of claims.

25. Smallpox inoculation was banned in England in 1840.

So how, then, do we model this in a task like NLI? We think that a potential answer to this question is to decompose a sentence into propositions, and to decompose each of the full propositions into all of its micro-propositions, before identifying the relation between two sentences. For example, the following micro-propositions could be derived from Sentence 25:

a. Smallpox inoculation was banned in England in 1840.
b. Smallpox inoculation was banned in England.
c. Smallpox inoculation was banned in 1840.
d. Smallpox inoculation was banned.
e. [something] was banned in England in 1840.
f. [something] was banned in England.
g. [something] was banned in 1840.
h. [something] was banned.

In turn, micro-propositions derived from other statements could be compared to each of the above, until we can pinpoint exactly which aspects of a given claim are contested, agreed upon, or left implicit. In addition to making our judgments more specific, we think that micro-propositions have the potential of making the NLI task more feasible on complex sentences. For instance, it would be quite hard for machines *and* humans to correctly determine the relation between the following two sentences without first identifying the perspectives toward specific aspects of the different claims made:

26. Pre-exposure vaccination is also recommended for individuals traveling to isolated areas or to areas where immediate access to appropriate medical care is limited or to countries where modern rabies vaccines are in short supply and locally available rabies vaccines might be unsafe and/or ineffective.
27. I'm sure that many others within the anti-vaccine movement have genuinely good intentions, and do honestly believe that vaccines are harmful.

Another reason why we believe that micro-propositions are important in perspective mining is that even within a sentence, only specific aspects of a proposition may be denied or confirmed. In the following subsections, we will review some of the research that has been done on this phenomenon in the context of factuality profiling and negation processing.

14.5.1 Micro-Propositions in Factuality Profiling

First, let us consider the relevance of micro-propositions for factuality profiling by an example. In FactBank, the factuality value assigned to the event *blowing (up)* in Sentence 28 below is PR+ ("did probably happen") for both relevant sources, namely the author and the non-explicit generic source introduced by the passive verb *suspected*. However, the uncertainty expressed by *suspected* is not so much directed toward the proposition as a whole as it is toward a specific aspect of the event: the two Libyans being the ones who did it. A more accurate representation of the factual status of this event would thus be to annotate BLOWING_UP(PAN AM JUMBO JET, OVER SCOTLAND, IN 1988) as CT+, and only BLOWING_UP(TWO LIBYANS) as PR+.

28. The World Court Friday rejected U.S. and British objections to a Libyan World Court case that has blocked the trial of two Libyans **suspected** of blowing up a Pan Am jumbo jet over Scotland in 1988.[8]

This phenomenon is also described in the Belief and Sentiment (BeSt) annotations,[9] which was one of the tracks in the 2016 and 2017 TAC-KBP evaluations (Ellis et al., 2016; Getman et al., 2017) and which annotations are based on the ERE (Entities, Relations, Events; Song et al., 2015) and the Committed Belief frameworks (Diab et al., 2009; Prabhakaran et al., 2015; Werner et al., 2015). In BeSt annotations, separate judgments are provided for both the event as a whole as for each of the event's arguments, acknowledging that the factual status of the argument's role in the event may differ from the event

[8] TimeBank/FactBank – APW19980227.476-S1.
[9] www.cs.columbia.edu/~rambow/best-eval-2016.

itself (Ellis et al., 2016). However, to our knowledge, neither reports on inter-annotator agreement to validate the quality of these annotations nor evaluation of system performance have been made available, and this phenomenon has not been captured in any of the other factuality corpora mentioned in this chapter. Therefore, it is still unclear whether it is feasible for humans and machines to reliably detect this fine-grained information in the context of factuality.

14.5.2 Micro-Propositions in Negation Processing

A related line of research is the annotation and detection of implicit positive meaning from negated statements. This task was pioneered by Blanco and Moldovan (2011), who argued that the interpretation of a negated statement depends on the location of *focus* in the sentence. For example, three possible locations of focus in Sentence 29 are *vaccines*, *polio* and *in the United States*, which would correspond to the following interpretations:

29. [Vaccines] **didn't** <u>eradicate</u> [polio] [in the Unites States].
    a. [something] eradicated polio in the United States, but not vaccines.
    b. Vaccines eradicated [something] in the United States, but not polio.
    c. Vaccines eradicated polio [somewhere], but not in the United States.

To capture this information, Blanco and Moldovan (2011) created the PB-FOC corpus, which contains annotations for the focus of negation in the 3,993 verbal negations in PropBank (Palmer et al., 2005). The highest system performance evaluated on PB-FOC was achieved by Zou et al. (2015) with a 69.3 accuracy using contextual discourse information.

In the last decade, PB-FOC has given rise to a couple of slight variations to the annotation of focus and definitions of the corresponding task. For instance, Anand and Martell (2012) reannotated the corpus by incorporating the pragmatic concept that discourse is guided by *questions under discussion*. Blanco and Moldovan (2012) introduced the annotation of fine-grained foci to capture even more specific interpretations in PB-FOC. Instead of annotating a single correct interpretation, Blanco and Sarabi (2016) scored all positive interpretations that could be derived from a negated statement by manipulating semantic roles according to their likelihood in a subset of OntoNotes (Hovy et al., 2006). Similar annotations were done by Sarabi and Blanco (2016), who scored more fine-grained positive interpretations by manipulating syntactic dependencies instead of semantic roles, and by Sanders and Blanco (2016), who applied the same method to modal constructions.

In terms of system performance, it seems like more annotated data is needed to be able to train and evaluate systems. Blanco and Sarabi (2016) report a Pearson's correlation of 0.64 on their Support Vector Machine (SVM) for regression. However, van Son et al. (2018) redefined the regression task proposed by Blanco and Sarabi (2016) as a classification task, with each positive interpretation to be classified as TRUE or FALSE (optionally, with a third class of UNCERTAIN representing the middle cases), and showed that a simple baseline that takes the mean over the scores or the most frequent class per semantic role in the training set achieves similar results. This was because of a strong class imbalance in the dataset, with too few examples of the low-frequency classes to be able to learn how to distinguish them.

14.6 Conclusions

In this chapter, we have reflected on perspectives from a natural language processing view. A perspective can be described as a relation between the source and the content of a statement. The relation is characterized by means of multiple perspective values characterizing the attitude of the source toward the statement. The perspective values can express sentiment, emotion, or factual level of commitment. We have argued that perspectives should be modeled at the micro-proposition level because the source can express attitudes toward different aspects of the propositions included in a statement.

We also have explained what the role of attribution and factuality is in the characterization of perspectives. Attribution allows us to link the source to the target of the propositional attitude, whereas factuality (further) specifies the level of commitment of the source toward this target. We have provided an overview of how attribution and factuality have been modeled for natural language processing and what are the main computational approaches for the automatic detection of these phenomena.

Finally, we have explained how the NLI task aims to tackle the next step of comparing claims and perspectives across texts. We argued that current datasets created for these tasks are not representative enough for the dispersed online debate that Web users face on their search for information; the sentences in these data sets are not as complex and varied as in a real setting, and the data often contains biases or an over-representation of explicit markers, allowing systems to perform well without actually having to analyze the content. But a great variety and complexity of natural online debate does present a challenge for propositional alignment; we have proposed a micro-propositional approach to make a (fine-grained) comparison of statements across texts feasible.

Most of the examples used in this chapter come from the Vaccination Corpus (Morante et al., 2020), which is a corpus of texts related to the online vaccination debate that so far has been annotated with vaccine-related events, attribution, claims, and opinions. One of our current efforts involves the annotation of factuality as well as opposition and agreement relations between automatically aligned propositions, taking into account the roles of sources and micro-propositions in these relations. We intend to use these annotations for the evaluation of both factuality and Natural Language Inference (NLI).

References

Almeida, M. S. C., Almeida, M. B., and Martins, A. F. T. 2014. A joint model for quotation attribution and coreference resolution. Pages 39–48 of: *Proceedings of the 14th Conference of the European Chapter of the Association for Computational Linguistics*. Gothenburg: Association for Computational Linguistics.

Anand, P., and Martell, C. 2012. Annotating the focus of negation in terms of questions under discussion. Pages 65–69 of: *Proceedings of the Workshop on Extra-Propositional Aspects of Meaning in Computational Linguistics*. Jeju, Republic of Korea: Association for Computational Linguistics.

Blanco, E., and Moldovan, D. 2011. Semantic representation of negation using focus detection. Pages 581–589 of: *Proceedings of the 49th Annual Meeting of the Association for Computational Linguistics: Human Language Technologies, Vol. 1*. Portland, Oregon: Association for Computational Linguistics.

Blanco, E., and Moldovan, D. 2012. Fine-grained focus for pinpointing positive implicit meaning from negated statements. Pages 456–465 of: *Proceedings of the 2012 Conference of the North American Chapter of the Association for Computational Linguistics: Human Language Technologies*. Montreal, Canada: Association for Computational Linguistics.

Blanco, E., and Sarabi, Z. 2016. Automatic generation and scoring of positive interpretations from negated statements. Pages 1431–1441 of: *Proceedings of the 2016 Conference of the North American Chapter of the Association for Computational Linguistics: Human Language Technologies*. San Diego: Association for Computational Linguistics.

Bowman, S. R., Angeli, G., Potts, C., and Manning, C. D. (2015), A large annotated corpus for learning natural language inference. Pages 632–642 of: Màquez, L., Callison-Burch, C., Su, J., Pighin, D., and Marton, Ys (eds.), *Empirical methods in natural language processing*. Lisbon, Portugal: The Association for Computational Linguistics.

Chen, S., Khashabi, D., Yin, W., Callison-Burch, C., and Roth, D. 2019. Seeing things from a different angle: Discovering diverse perspectives about claims. Pages 542–557 of: *Proceedings of the 2019 Conference of the North American Chapter of the Association for Computational Linguistics: Human Language Technologies, Vol. 1: Long and Short Papers*. Minneapolis, MN: Association for Computational Linguistics.

Dagan, I., Glickman, O., and Magnini, B. 2006. The PASCAL recognising textual entailment challenge. *Lecture Notes in Computer Science (Including Subseries Lecture Notes in Artificial Intelligence and Lecture Notes in Bioinformatics)*, **3944 LNAI**, 177–190.

de La Clergerie, É., Sagot, B., Stern, R., Denis, P., Recourcé, G., and Mignot, V. 2011. Extracting and visualizing quotations from news wires. Pages 522–532 of: *Human language technology. Challenges for computer science and linguistics*. Springer.

De Marneffe, M.-C., Manning, C. D., and Potts, C. 2012. Did it happen? The pragmatic complexity of veridicality assessment. *Computational Linguistics*, **38**(2), 301–333.

Diab, M. T., Levin, L., Mitamura, T., Rambow, O., Prabhakaran, V., and Guo, W. 2009. Committed belief annotation and tagging. Pages 68–73 of: *Proceedings of the Third Linguistic Annotation Workshop (Association for Computational Linguistics-International Joint Conference on Natural Language Processing 2009)*.

Ellis, J., Getman, J., Kuster, N., Song, Z., Bies, A., and Strassel, S. 2016. Overview of linguistic resources for the TAC KBP 2016 evaluations: Methodologies and results. Technical report. Linguistic Data Consortium, University of Pennsylvania.

Elson, D. K., and McKeown, K. 2010. Automatic attribution of quoted speech in literary narrative. Page 201 of: *Twenty-Fourth Association for the Advancement of Artificial Intelligence Conference on Artificial Intelligence*.

Getman, J., Ellis, J., Song, Z., Tracey, J., and Strassel, S. 2017. Overview of linguistic resources for the TAC KBP 2017 evaluations: Methodologies and results. Technical report. Linguistic Data Consortium, University of Pennsylvania.

Gururangan, S., Swayamdipta, S., Levy, O., Schwartz, R., Bowman, S. R., and Smith, N. A. 2018. Annotation artifacts in natural language inference data. Pages 107–112 of: *Proceedings of the 2018 Conference of the North American Chapter of the Association for Computational Linguistics: Human Language Technologies*. Location: New Orleans, Louisiana. Publisher: Association for Computational Linguistics.

Hovy, E., Marcus, M., Palmer, M., Ramshaw, L., and Weischedel, R. 2006. OntoNotes: The 90% solution. Pages 57–60 of: *Proceedings of the 2006 Conference of the North American Chapter of the Association for Computational Linguistics: Human Language Technologies*. Location: New York City, USA. Publisher: Association for Computational Linguistics, Companion Volume: Short Papers.

Khot, T., Sabharwal, A., and Clark, P. 2018. SCITAIL: A textual entailment dataset from science question answering. Pages 5189–5197 of: *Association for the Advancement of Artificial Intelligence Conference on Artificial Intelligence Conference on Artificial Intelligence*.

Lai, A., Bisk, Y., and Hockenmaier, J. 2017. Natural language inference from multiple premises. Pages 100–109 of: *Proceedings of the Eighth International Joint Conference on Natural Language Processing, Vol. 1: Long Papers*. Taipei: Asian Federation of Natural Language Processing.

Lee, K., Artzi, Y., Choi, Y., and Zettlemoyer, L. 2015. Event detection and factuality assessment with non-expert supervision. Pages 1643–1648 of: *Proceedings of the 2015 Conference on Empirical Methods in Natural Language Processing*.

Lotan, A., Stern, A., and Dagan, I. 2013. TruthTeller: Annotating predicate truth. Pages 752–757 of: *Proceedings of the 2013 Conference of the North American Chapter of the Association for Computational Linguistics: Human Language Technologies*.

Marelli, M., Menini, S., Baroni, M., Bentivogli, L., Bernardi, R., and Zamparelli, R. 2014. A SICK cure for the evaluation of compositional distributional semantic models. Pages 216–223 of: *Proceedings of the Ninth International Conference on Language Resources and Evaluation*. LREC'14. Reykjavik: European Language Resources Association.

Minard, A.-L., Speranza, M., Urizar, R., Altuna, B., van Erp, M., Schoen, A., and van Son, C. 2016. MEANTIME, the NewsReader multilingual event and time corpus. Pages 4417–4422 of: *Proceedings of the Tenth International Conference on Language Resources and Evaluation*. LREC '16. Portorož: European Language Resources Association.

Morante, R., van Son, C., Maks, I., and Vossen, P. 2020. Annotating perspectives on vaccination. Pages 4964–4973 of: *Proceedings of the 12th Language Resources and Evaluation Conference*. LREC '20, Marseille: European Language Resources Association.

Muzny, G., Fang, M., Chang, A. X., and Jurafsky, D. 2017. A two-stage sieve approach for quote attribution. Pages 460–470 of: *15th Conference of the European Chapter of the Association for Computational Linguistics, EACL 2017 - Proceedings of Conference, Vol. 1: Long Papers*. Valencia: Association for Computational Linguistics.

Nangia, N., Williams, A., Lazaridou, A., and Bowman, S. 2017. The RepEval 2017 shared task: Multi-genre natural language inference with sentence representations. Pages 1–10 of: *Proceedings of the 2nd Workshop on Evaluating Vector Space Representations for NLP*, Copenhagen: Association for Computational Linguistics.

Newell, E., Margolin, D., and Ruths, D. 2018. An attribution relations corpus for political news. *Proceedings of the 11th Language Resources and Evaluation Conference*. Miyazaki: European Language Resource Association.

O'Keefe, T., Pareti, S., Curran, J. R., Koprinska, I., and Honnibal, M. 2012. A sequence labelling approach to quote attribution. Pages 790–799 of: *Proceedings of the 2012 Joint Conference on Empirical Methods in Natural Language Processing and Computational Natural Language Learning*. Jeju Island: Association for Computational Linguistics.

Palmer, M., Gildea, D., and Kingsbury, P. 2005. The proposition bank: A corpus annotated with semantic roles. *Computational Linguistics Journal*, **31**(1), 71–106.

Pareti, S. 2012. The independent encoding of attribution relations. Pages 48–55 of: *Proceedings of the Eight Joint ACL-ISO Workshop on Interoperable Semantic Annotation (ISA-8)*.

Pareti, S. 2015. Attribution: A computational approach. Ph.D. thesis, University of Edinburgh.

Pareti, S., O'Keefe, T., Konstas, I., Curran, J. R., and Koprinska, I. 2013. Automatically detecting and attributing indirect quotations. Pages 989–999 of: *Proceedings of the 2013 Conference on Empirical Methods in Natural Language Processing*. Seattle, WA: Association for Computational Linguistics.

Poliak, A., Haldar, A., Rudinger, R., Hu, J. E., Pavlick, E., White, A. S., and Van Durme, B. 2018a. Collecting diverse natural language inference problems for sentence representation evaluation. Pages 67–81 of: *Proceedings of the 2018 Conference on Empirical Methods in Natural Language Processing*. Brussels: Association for Computational Linguistics.

Poliak, A., Naradowsky, J., Haldar, A., Rudinger, R., and Van Durme, B. 2018b. Hypothesis only baselines in natural language inference. Pages 180–191 of: *Proceedings of the 7th Joint Conference on Lexical and Computational Semantics*. *SEM '18, New Orleans, Louisiana: Association for Computational Linguistics.

Prabhakaran, V., By, T., Hirschberg, J., Rambow, O., Shaikh, S., Strzalkowski, T., Tracey, J., Arrigo, M., Basu, R., Clark, M., Dalton, A., Diab, M., Guthrie, L., Prokofieva, A., Strassel, S., Werner, G., Wiebe, J., and Wilks, Y. 2015. A new dataset and evaluation for belief/factuality. Pages 82–91 of: *Proceedings of the Fourth Joint Conference on Lexical and Computational Semantics*. *SEM '15. Denver, CO: Association for Computational Linguistics.

Prabhakaran, V., Rambow, O., and Diab, M. 2010. Automatic committed belief tagging. Pages 1014–1022 of: *Proceedings of the 23rd International Conference on Computational Linguistics. COLING '10: Posters*. Beijing: Association for Computational Linguistics.

Prasad, R., Dinesh, N., Lee, A., Joshi, A., and Webber, B. 2007. Attribution and its annotation in the Penn Discourse TreeBank. *Traitement Automatique des Langues, Special Issue on Computational Approaches to Document and Discourse*, **47**(2), 43–64.

Rudinger, R., White, A. S., and Van Durme, B. 2018. Neural models of factuality. Pages 731–744 of: *Proceedings of the 2018 Conference of the North American Chapter of the Association for Computational Linguistics: Human Language Technologies, Vol. 1: Long Papers*. New Orleans, LA: Association for Computational Linguistics.

Sanders, J., and Blanco, E. 2016. Automatic extraction of implicit interpretations from modal constructions. Pages 1098–1107 of: *Proceedings of the 2016 Conference on Empirical Methods in Natural Language Processing*. EMNLP '16. Austin, Texas. Association for Computational Linguistics.

Sarabi, Z., and Blanco, E. 2016. Understanding negation in positive terms using syntactic dependencies. Pages 1108–1118 of: *Proceedings of the 2016 Conference on Empirical Methods in Natural Language Processing*. EMNLP '16. Austin, Texas: Association for Computational Linguistics.

Saurí, R. 2008. A factuality profiler for eventualities in text. Ph.D. thesis, Brandeis University, Waltham, MA.

Saurí, R., and Pustejovsky, J. 2009. FactBank: A corpus annotated with event factuality. *Language resources and evaluation*, **43**(3), 227–268.

Scheible, C., Klinger, R., and Padó, S. 2016. Model architectures for quotation detection. Pages 1736–1745 of: *Proceedings of the 54th Annual Meeting of the Association for Computational Linguistics, ACL 2016, Vol. 3: Long Papers*, Berlin: Association for Computational Linguistics.

Schneider, N., Hwa, R., Gianfortoni, P., Das, D., Heilman, M., Black, A. W., Crabbe, F. L., and Smith, N. A. 2010. Visualizing topical quotations over time to understand news discourse. Technical report. CMU-LTI-10-013. Carnegie Mellon University, Pittsburgh, PA.

Song, Z., Bies, A., Strassel, S., Riese, T., Mott, J., Ellis, J., Wright, J., Kulick, S., Ryant, N., and Ma, X. 2015. From light to rich ERE: Annotation of entities, relations, and events. Pages 89–98 of: *Proceedings of the 3rd Workshop on EVENTS at the North American Chapter of the Association for Computational Linguistics: Human Language Technologies*. Location: Denver, Colorado. Association for Computational Linguistics.

Stanovsky, G., Eckle-Kohler, J., Puzikov, Y., Dagan, I., and Gurevych, I. 2017. Integrating deep linguistic features in factuality prediction over unified datasets. Pages 352–357 of: *Proceedings of the 55th Annual Meeting of the Association for Computational Linguistics, Vol. 2: Short Papers*, Vancouver: Association for Computational Linguistics.

Tsuchiya, M. 2018. Performance impact caused by hidden bias of training data for recognizing textual entailment. Pages 1506–1511 of: *Proceedings of the 11th International Conference on Language Resources and Evaluation* (LREC 2018). Miyazaki, Japan: European Language Resources Association (ELRA).

van Son, C., Morante, R., Aroyo, L., and Vossen, P. 2018. Scoring and classifying implicit positive interpretations: A challenge of class imbalance. *Proceedings of the 27th International Conference on Computational Linguistics*. COLING '18, Santa Fe, New Mexico: Association for Computational Linguistics.

van Son, C., van Erp, M., Fokkens, A., and Vossen, P. 2014. Hope and fear: Interpreting perspectives by integrating sentiment and event factuality. Pages 26–31 of: *Proceedings of the 9th International Conference on Language Resources and Evaluation*. LREC '14, Reykjavik: European Language Resources Association.

Vilares, D., and He, Y. 2017. Detecting perspectives in political debates. Pages 1573–1582 of: *Proceedings of the 2017 Conference on Empirical Methods in Natural Language Processing*. Copenhagen: Association for Computational Linguistics.

Weiser, S., and Watrin, P. 2012. Extraction of unmarked quotations in newspapers. Pages 559–562 of: Calzolari, N., Choukri, K., Declerck, T., Dogan, M. U., Maegaard, B., Mariani, J., Odijk, J., and Piperidis, S. (eds), *Proceedings of the 8th International Conference on Language Resources and Evaluation, LREC 2012*. Istanbul, Turkey: European Language Resources Association.

Werner, G. J., Prabhakaran, V., Diab, M., and Rambow, O. 2015. Committed belief tagging on the FactBank and LU corpora: A comparative study. Pages 32–40 of: *Proceedings of the Second Workshop on Extra-Propositional Aspects of Meaning in Computational Semantics*. ExProM '15, Denver, Colorado: Association for Computational Linguistics.

White, A. S., Reisinger, D., Sakaguchi, K., Vieira, T., Zhang, S., Rudinger, R., Rawlins, K., and Van Durme, B. 2016. Universal decompositional semantics on universal dependencies. Pages 1713–1723 of: *Proceedings of the 2016 Conference on Empirical Methods in Natural Language Processing*. Austin, TX: Association for Computational Linguistics.

Zou, B., Zhu, Q., and Guodong, Z. 2015. Unsupervised negation focus identification with word-topic graph model. Pages 1632–1636 of: *Proceedings of the 2015 Conference on Empirical Methods in Natural Language Processing*. EMNLP '15, Lisbon, Portugal: Association for Computational Linguistics.

15
Toward Automatic Discovery of Diverse Perspectives

Sihao Chen, Daniel Khashabi, and Dan Roth

15.1 Introduction

Understanding most nontrivial *claim*s requires insights from various *perspectives*. Today, we make use of search engines or recommendation systems to retrieve information relevant to a claim. Such systems serve information with personalized preferences, but often at the cost of decreasing the diversity of views presented (Fletcher and Nielsen, 2018). For example, the recommendations are often optimized for maximizing the popularity of the relevant content, rather than reflecting the variety of the authors' or sources' *perspective*s presented in them or whether they are supported by evidence (Steiner et al., 2020).

In this chapter, we explore a line of work on mitigating such *selection bias* (Heckman, 1979) when retrieving perspectives with respect to a (disputed) claim. Consider the *claim* shown in Figure 15.1: *"animals should have lawful rights."* One might compare the biological similarities/differences between humans and other animals to support/oppose the claim. Alternatively, one can base an argument on morality and rationality of animals, or lack thereof. Each of these arguments, which we will refer to as *perspective*s throughout the chapter, are opinions, possibly conditional, in support of a given *claim* or against it. Each unique *perspective* thus constitutes a particular attitude toward a given *claim*.

Natural language understanding is at the heart of developing an ability to identify diverse perspectives for claims. We propose and study a setting that would facilitate discovering *diverse perspectives* and their supporting evidence with respect to a given *claim*. Our goal is to identify and formulate the key NLP challenges underlying this task and develop a dataset that would allow a systematic study of these challenges. For example, for the claim in Figure 15.1, multiple (non-redundant) perspectives should be retrieved from a pool

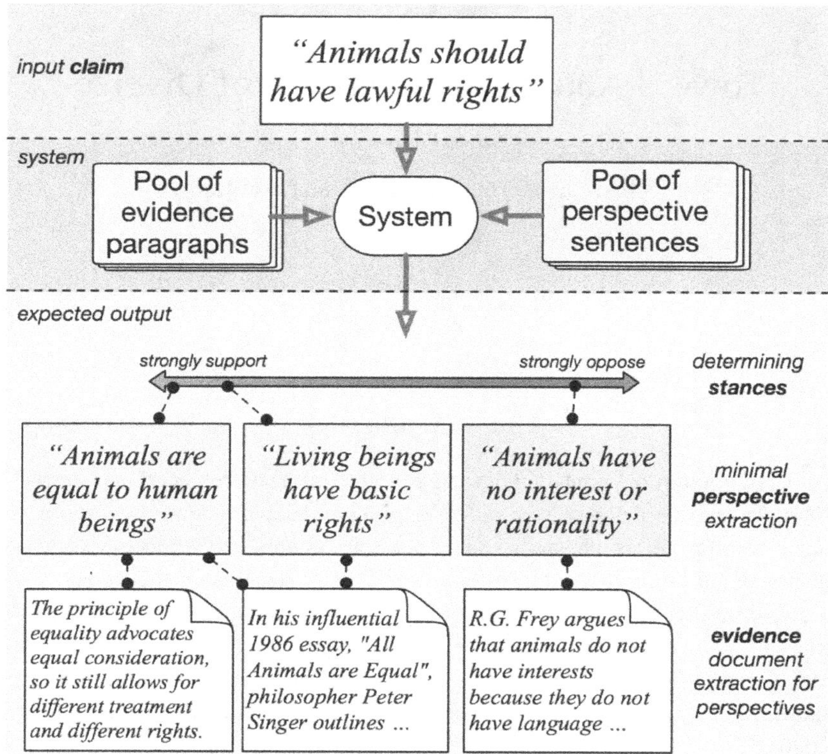

Figure 15.1 Given a *claim*, a hypothetical system is expected to discover various *perspectives* that are substantiated with *evidence* and their *stance* with respect to the claim

of perspectives; one of them is *"animals have no interest or rationality,"* a *perspective* that should be identified as taking an *opposing* stance with respect to the *claim*. Each *perspective* should also be well supported by *evidence* found in a pool of potential pieces of evidence. While it might be impractical to provide an exhaustive spectrum of ideas with respect to a *claim*, presenting a small but diverse set of *perspectives* could be an important step toward addressing the *selection bias* problem. Moreover, it would be impractical to develop an exhaustive pool of evidence for all perspectives from a diverse set of credible sources, and we are not attempting to do that. We aim at formulating the core NLP problems and developing a dataset of diverse yet not necessarily exhaustive perspectives to facilitate the study of these problems. Inherently, our objective requires understanding of the relations between *perspective*s and *claim*s, the semantic nuances within various *perspective*s in the context of *claims*, and relations between perspectives and evidence. We leave out the trustworthiness of perspectives and evidence paragraph in our setting, as we argue that the credibility of sources should be studied as an equally important but orthogonal problem.

In summary, this chapter is dedicated to a series of works[1] toward automatic discovery of *diverse perspectives* with respect to a given *claim*.

- As a starting point, we identify and characterize the core NLP challenges required to solve the *perspective discovery* problem. To consolidate our task formulation and facilitate research in this direction, we construct *Perspectrum*,[2] a dataset of *claims*, *perspectives*, and *evidence* paragraphs. In this dataset, for a given *claim* and pools of *perspectives* and *evidence paragraphs*, we expect a system to select the relevant perspectives and their supporting paragraphs.
- Based on the annotated dataset, we develop competitive systems for each sub-task, using state-of-the-art techniques. By assembling the learned individual components, we build PERSPECTROSCOPE 🔍,[3] a platform that simulates end-to-end process of minimal perspective discovery.

15.2 Minimal Perspective Discovery: Tasks and Challenges

In this section, we take a closer look into the challenge and define a collection of tasks that move us closer to *diverse perspective discovery*. For brevity, we use the following notations throughout the rest of the chapter. Let c be a target claim of interest (for example, the claims c_1 and c_2 in Figure 15.2). Each claim c is addressed by a collection of perspectives $\{p\}$ that are grouped into clusters of *equivalent* perspectives. Additionally, each perspective p to a claim c can be supported by a set of evidence paragraphs $\{e\}$, denoted $e \vDash p|c$.

Creating systems that would address the challenges in full glory requires solving the following interdependent tasks:

(T0) *Identifying debate-worthy claims:* not every claim requires an in-depth discussion of perspectives. For a system to be practical, it needs to be equipped with the ability to understand argumentation structures (Palau and Moens, 2009) in order to discern disputed claims from those with straightforward

[1] Earlier versions of this work appeared in Chen et al. (2019b,a).

[2] `https://github.com/CogComp/perspectrum.`

[3] `http://perspectroscope.com/.`

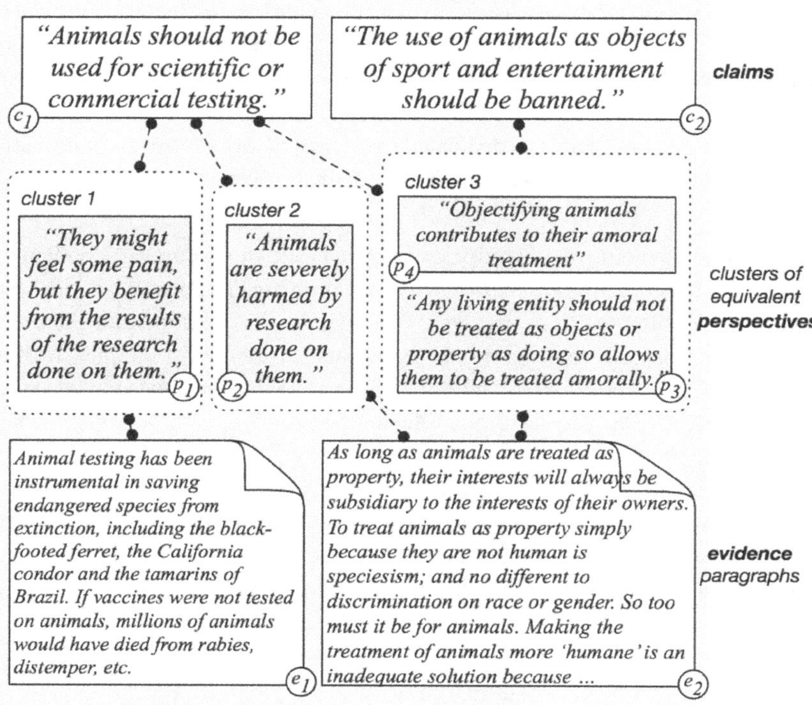

Figure 15.2 Depiction of a few claims, their *perspectives*, and evidences from Perspectrum. The *opposing* and *supporting* perspectives are indicated with dark (left side box) and light (middle and right side boxes), respectively

responses. We set aside this problem in our study and assume that all the inputs to the systems are discussion-worthy claims.

***(T1)** Discovery of pertinent perspectives:* a system is expected to recognize argumentative sentences (Cabrio and Villata, 2012) that directly address the points raised in the disputed claim. For example, in Figure 15.2 while all the perspectives are topically related to all the claims, p_1, p_2 do not directly address the focus of claim c_2 (i.e., *"use of animals" in "entertainment"*). Inherently, this objective involves reasoning about the core arguments provoked by a claim or a perspective and thus requiring deeper language and knowledge understanding than mere topic matching.

***(T2)** Perspective equivalence:* a system is expected to extract a *minimal* and *diverse* set of perspectives. This requires the ability to decide equivalent perspectives p and p' with respect to a claim c: $p|c \approx p'|c$. For instance, in Figure 15.2 p_3 and p_4 are equivalent in the context of c_2; however, they might not be equivalent with respect to any of the claims. The conditional nature of perspective equivalence differentiates it from the *paraphrasing* task (Bannard and Callison-Burch, 2005).

(T3) *Stance classification of perspectives:* a system is supposed to assess the stances of the perspectives with respect to the given claim (supporting, opposing, etc.) (Hasan and Ng, 2014).

(T4) *Substantiating the perspectives:* a system is expected to find valid evidence paragraph(s) in support of each perspective. Conceptually, this is similar to the well-studied problem of textual entailment (Dagan et al., 2013), except that here the entailment decisions depend on the choice of claims.

15.3 Related Work

Claim Verification. The task of *fact verification* or *fact-checking* focuses on the assessment of the truthfulness of a claim, given evidence (Karimi et al., 2018; Mitra and Gilbert, 2015; Samadi et al., 2016; Vlachos and Riedel, 2014; Wang, 2017). These tasks are highly related to the task of textual entailment that has been extensively studied in the field (Bentivogli et al., 2008; Dagan et al., 2013). Some recent work studies jointly the problem of identifying evidence and verifying that it supports the claim (Yin and Roth, 2018). In a similar vein, our problem structure encompasses the *fact verification* problem (as verification of *perspectives* from *evidence*; Figure 15.1).

Stance Classification. Stance classification aims at detecting phrases that *support* or *oppose* a given claim. The problem has gained significant attention in recent years; to note a few important ones, Hasan and Ng (2014) create a dataset of text snippets, annotated with finite "reasons" (similar to *perspectives* in this work) and stances (whether they support or oppose the claim). Ferreira and Vlachos (2016) create a dataset of rumors (claims) coupled with news headlines and their stances. A few notable works that belong to this category are Bar-Haim et al. (2017), Mohammad et al. (2016), Park and Cardie (2014), Rinott et al. (2015), and Sobhani et al. (2017). Our approach here is closely related to existing work in this direction, as stance classification is an essential element of the problem studied here. The definitions of these problems are equivalent to our ***(T3)*** sub-task formulation; however, the existing datasets target different domains (e.g., topics, sources of claims, etc.) than our work.

Perspective Discovery. A few recent datasets use a similar conceptual design that involves a *claim*, *perspectives*, and *evidence*. These works are either too small due to the high cost of construction (Aharoni et al., 2014) or too noisy because of the way they are crawled from online resources (Hua and Wang, 2017; Wachsmuth et al., 2017a). In order to construct a sizable and high-quality dataset, we make use of a mix of online content and crowdsourcing. Existing online content is much cheaper to obtain than building a dataset from scratch. However, since they are noisy, we used crowdsourcing to remove the erroneous content.

There exist a few related efforts among other online platforms. `args.me` is a platform that accepts natural language queries and returns links to the pages that contain relevant topics (Wachsmuth et al., 2017b), which are split into *supporting* and *opposing* categories. Similarly, `ArgumentText` (Stab et al., 2018) takes a topic as input and returns *pro/con* arguments retrieved from the Web. This work takes these efforts further by employing language understanding techniques.

15.4 The *Perspectrum* Dataset

15.4.1 Dataset Construction

This section describes the construction of *Perspectrum*. The core of this process involves crowdsourcing using Amazon Mechanical Turk (AMT). Due to the complexity of the target task, the annotation process was divided into multiple distinct steps, as outlined in Figure 15.3.

For any of the annotation steps, crowd workers are guided to an external platform where they are required to first read the instructions and pass a qualification test to make sure they have understood the instructions. Crowd workers are allowed to start the annotation tasks only after successful completion of the test. More details on the dataset construction can be found in the extended version of this work (Chen et al., 2019a).

(1) Initial Data Collection. We start by crawling the content of a few notable debate websites: `idebate.com, debatewise.org, procon.org.`

(2.1) Perspective Verification. For each perspective, we annotate (a) its grammaticality (whether the perspective is a proper English sentence or not), (b) whether it is taking a clear stance with respect to the given claim. Only those perspectives that satisfy these properties (with a reasonable level of agreement) are retained.

(2.2) Perspective Paraphrases. To enrich the ways the perspectives are phrased, we ask crowd workers to generate multiple paraphrases for each

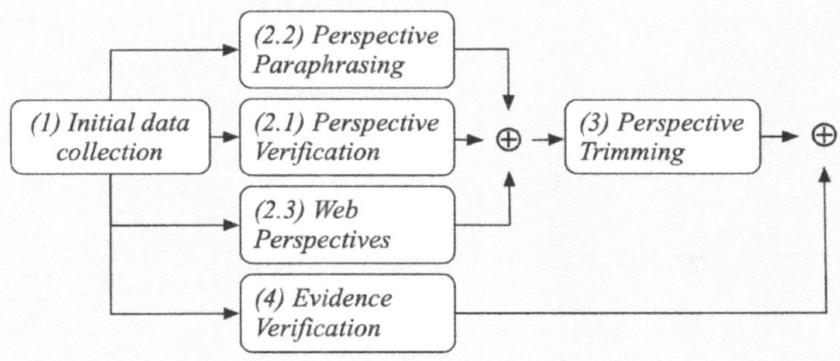

Figure 15.3 The pipeline of dataset construction for *Perspectrum*

perspective. We then verify the quality in a second-round annotation and retain the high-quality ones.

(2.3) Web Perspectives. To include candidate perspectives with more natural phrasings, we use Web search to retrieve sentences from the Web. We annotate them with the appropriate relevance and stance labels and retain the ones that are agreed upon.

(3) Final Perspective Trimming. An expert annotator goes over all the claims in order to verify that (a) all the equivalent perspectives are clustered together; (b) perspectives are properly shared between the two similar claims.

(4) Evidence Verification. A list of most-similar perspective-evidence pairs are annotated by crowd workers to determine whether the paragraph *supports* a given perspective or not. Those perspectives with reasonable agreement are retained.

15.4.2 Notable Statistics on *Perspectrum*

The dataset contains about $1k$ claims with a significant length diversity. Additionally, the dataset comes with $\sim 12k$ perspectives. The perspectives which convey the same point with respect to a claim are grouped into clusters. On average, each cluster has a size of 2.3, which shows that, on average, many perspectives have equivalents.

15.5 Building Systems Based on *Perspectrum*

We provide empirical analysis to learn the tasks. To evaluate the generalization, we create a split of 60%/15%/25% of the data train/dev/test. In order to make

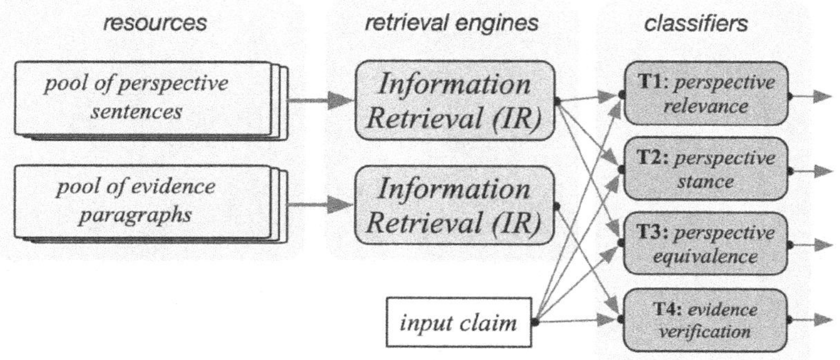

Figure 15.4 A high-level diagram showing the connection between the classifiers trained for the different tasks and other resources

sure that our systems are generalizing across topics, we make sure claims with the same topic fall into the same split.
We make use of the following systems in our evaluation:

Information Retrieval. This baseline has been successfully used for related tasks like Question Answering (Clark et al., 2016). We use this system to retrieve a ranked list of best matching perspective/evidence from the corresponding pool of documents.

Pre-trained Contextual Models. We consider two recent systems in this class. BERT is a recent state-of-the-art contextualized representation (Devlin et al., 2018). RoBERTa is another state-of-the-art contextualized representation (Liu et al., 2019) that is built by extending BERT. This system has shown significantly good performance on language understanding tasks (Clark et al., 2019). For both of these systems, we consider two variations (*large* and *base*) that are different in their parameter size.

Human Performance. Human performance provides us with an estimate of the best achievable results on datasets.

15.5.1 In-Domain Evaluations

We train separate models for each of the four different sub-tasks ***T1–T4***, and evaluate them on their corresponding test splits. A high-level picture of the decision-making pipeline is shown in Figure 15.4. For each input claim, first, a retrieval system is expected to generate candidates from two given pools

Table 15.1 *Quality of different baselines on different sub-tasks (Section 15.5). All the numbers are in percentage. Top machine baselines are in* **bold**

Setting	Task	System	*Pre.*	*Rec.*	*F*1
T1: Perspective Relevance	Given a claim, select a list of relevant perspectives from the pool of documents.	IR	46.8	34.9	40.0
		IR + BERT$_{base}$	47.3	54.8	50.8
		IR + BERT$_{large}$	72.9	43.9	54.8
		IR + RoBERTa$_{base}$	66.4	49.3	56.6
		IR + RoBERTa$_{large}$	88.2	53.4	**66.4**
		IR + Human	63.8	83.8	72.5
T2: Perspective Stance	Given a claim, and collection of (gold) relevant perspectives, decide their stances.	Always "supp."	51.6	100.0	68.0
		BERT$_{base}$	70.5	71.1	70.8
		BERT$_{large}$	75.2	84.4	79.6
		RoBERTa$_{base}$	83.8	86.8	85.3
		RoBERTa$_{large}$	90.5	92.8	**91.6**
		Human	91.3	90.6	90.9
T3: Perspective Equivalence	Given a claim, and collection of (gold) relevant perspectives, decide which pairs are equivalent.	Always "equiv."	20.3	100.0	33.7
		IR	36.5	36.5	36.5
		BERT$_{base}$	85.3	50.8	63.7
		BERT$_{large}$	77.0	80.9	78.9
		RoBERTa$_{base}$	71.3	89.1	79.2
		RoBERTa$_{large}$	80.6	88.6	**84.4**
		Human	87.5	80.2	83.7
T4: Evidence Extraction	Given a claim and a perspective, find their list of evidence paragraphs from the pool of evidence documents.	IR	42.2	52.5	46.8
		IR + BERT$_{base}$	69.7	46.3	55.7
		IR + BERT$_{large}$	77.7	48.9	60.0
		IR + RoBERTa$_{base}$	69.4	51.4	59.4
		IR + RoBERTa$_{large}$	73.4	56.5	**63.8**
		IR + Human	70.8	53.1	60.7

of documents (one for perspectives and one for evidence documents). The candidates are then passed down to four classifiers trained for the individual sub-tasks. The overall results are summarized in Table 15.1. Each task is evaluated as a binary prediction task, and we use the standard definitions for Precision, Recall, and F1 metrics for our evaluations.

From the empirical results, a couple of observations that stand out are as follows.

(a) In almost all cases, the retrieval baselines lag considerably behind the human score. We also observe recall bottlenecks for perspective and evidence candidate generation in tasks ***T1*** and ***T4***, indicating that retrieval-based solution alone is not capable of recognizing the nuanced relation among claims, perspectives, and evidence paragraphs.
(b) Pretrained contextual language models are very effective across all tasks. The best-performing model is RoBERTa$_{large}$, which matches human performance on our benchmark and based on our evaluations on ***T2–T4***. We believe that the remarkable gains in RoBERTa (compared to BERT) are made possible by more extensive pre-training on significantly larger textual data (Liu et al., 2019). While the contextual language models yield optimistic results on our dataset, it remains to be seen whether the learned systems are directly applicable for perspective discovery on a broader scope, as we explore in later sections.

15.5.2 Out-of-Domain Generalization

Generalization to Unseen Topics. To better understand the generalization capability of our models to unseen topics, we collect claims from various Web sources (such as newswire and forums). We sample fifty random claims from (i) debate forums and (ii) Google search auto-complete. We first use Google to retrieve relevant pages and take the sentences in the retrieved pages as candidates to our classifiers. Among the most relevant results (scored by the best model for ***T1***), we manually evaluate top fourteen perspectives. Two authors independently label the perspectives with appropriate relevance labels and adjudicate their discrepancies. We evaluate the best model on the manually labeled data, and it scores 43.7 and 72.2 F_1, respectively on the debate forum claims and Google auto-complete claims. This finding suggests the need to build more comprehensive evaluations in order to gain more realistic understanding of our progress on the perspective discovery task.

Robustness to Local Modifications. To better understand the robustness of our models, we evaluate on modified but nontrivial versions of our claims and perspectives (Gardner et al., 2020). These annotations are constructed via local modifications on the claims by the authors in the following steps. First, we created nontrivial negations of the claim, for example, *Should we live in space?* $\rightarrow$ *Should we drop the ambition to live in space?* Next, we verify the stance of the edited claim with respect to each perspective. Overall, the annotators create 217 modified instances. Finally, another reviewer checks and adjudicates the annotations to ensure their quality. By evaluating RoBERTa$_{large}$ on the modified instances, we observe a drop of 4.6% F_1 in the model performance, indicating the model's sensitivity to minor input changes.

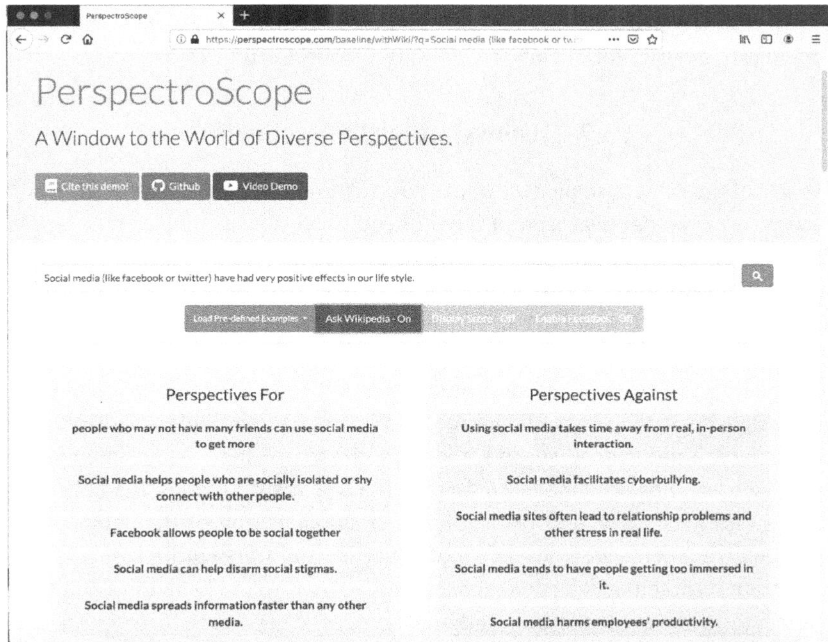

Figure 15.5 Given a *claim* as the input, our system is expected to discover various *perspectives* and their *stance* with respect to the claim. Each claim also comes with the relevant *evidence* that substantiates the given perspective

15.6 Toward Automated Perspective Discovery

Toward the conceptual framework of automatic perspective discovery (Section 15.1), we present PERSPECTROSCOPE 🔍, which let users explore different perspectives related to *any* natural-language input claim.

The high-level structure of the system resembles the decision-making pipeline shown in Figure 15.4. Our system receives a *claim* and is expected to identify a *diverse* set of *well-corroborated perspectives* that take a *stance* with respect to the claim. Each perspective should be substantiated by *evidence* paragraphs that summarize pertinent results and facts (a screenshot in Figure 15.5).

In contrast to our training settings, we extend the sources of perspectives and evidence paragraphs to a few well-known opinion/news websites like Washington Post. Additionally, we use the learned classifiers (Section 15.5) to rank and categorize each perspective based on their relevance and stance, as shown in Figure 15.5. In addition, users have the option to submit feedback to each perspective based on the quality/correctness of the predictions. Such feedback can potentially be used for evaluation, as suggested in Section 15.5.2.

15.7 Discussion and Conclusion

As one of the key consequences of the information revolution, *information pollution* and *over-personalization* have already had detrimental effects on our life. In this work, we attempt to facilitate the development of systems that aid in better organization and access to information, with the hope that the access to more diverse information can address over-personalization too (Vydiswaran et al., 2014).

There are a few aspects of the problem that we defer to future studies. First, we assume that the input claims to the system are reasonably well-phrased and indeed opinionated. Second, the work presented here is not intended to be *exhaustive*, nor does it attempt to reflect a true distribution of the important claims and perspectives in the world or to associate any of the perspectives and identified evidence with levels of expertise and trustworthiness. It is also important to note that when we ask crowd workers to evaluate the validity of perspectives and evidence, their judgment process can potentially be influenced by their prior beliefs. To avoid additional biases introduced in the process of dataset construction, we try to take the least restrictive approach in filtering dataset content beyond the necessary quality assurances. For this reason, we choose not to explicitly ask annotators to filter content based on the intention of its creators (e.g., offensive content). In a similar vein, since our main focus is the study of the relations between *claim*s, *perspective*s, and *evidence*, we leave out important issues such as their degree of factuality (Vlachos and Riedel, 2014) or trustworthiness (Pasternack and Roth, 2010, 2014) as separate aspects of the problem. We hope that some of these challenges and limitations will be addressed in future work.

In summary, we define the problem of *substantiated perspective discovery* and characterize language understanding tasks necessary to address this problem. We combine online resources, Web data, and crowdsourcing and create a high-quality dataset in order to drive research on this problem. Finally, we build and evaluate strong baseline supervised systems for this problem. We present a powerful interface for exploring different perspectives to discussion-worthy claims. The system is built with a combination of retrieval engines and learned classifiers to create a good balance between speed and quality. Our system is designed with the mindset of being able to get feedback from users of the system. We hope that our study will motivate further research in this direction.

References

Aharoni, E., Polnarov, A., Lavee, T., Hershcovich, D., Levy, R., Rinott, R., Gutfreund, D., and Slonim, N. 2014. A benchmark dataset for automatic detection of claims and evidence in the context of controversial topics. Pages 64–68 of: *Proceedings of the Workshop on Argumentation Mining*. Baltimore, Maryland: Association for Computational Linguistics.

Bannard, C. and Callison-Burch, C. 2005. Paraphrasing with bilingual parallel corpora. Pages 597–604 of: *Proceedings of Association for Computational Linguistics*.

Bar-Haim, R., Bhattacharya, I., Dinuzzo, F., Saha, A., and Slonim, N. 2017. Stance classification of context-dependent claims. Pages 251–261 of: *Proceedings of EACL*.

Bentivogli, L., Clark, P., Dagan, I., and Giampiccolo, D. 2008. The sixth PASCAL recognizing textual entailment challenge. Pages of 177–190 *TAC*.

Cabrio, E. and Villata, S. 2012. Combining textual entailment and argumentation theory for supporting online debates interactions. Pages 208–212 of: *Proceedings of ACL*.

Chen, S., Khashabi, D., Callison-Burch, C., and Roth, D. 2019a. PerspectroScope: A window to the world of diverse perspectives. Pages 129–134 of: *Proceedings of ACL Demonstrations*.

Chen, S., Khashabi, D., Yin, W., Callison-Burch, C., and Roth, D. 2019b. Seeing things from a different angle: Discovering diverse perspectives about claims. Pages 542–557 of: *Proceedings of the 2019 Conference of the North American Chapter of the Association for Computational Linguistics: Human Language Technologies*.

Clark, P., Etzioni, O., Khashabi, D., Khot, T., Mishra, B. D., Richardson, K., Sabharwal, A., Schoenick, C., Tafjord, O., Tandon, N., Bhakthavatsalam, S., Groeneveld, D., Guerquin, M., and Schmitz, M. 2019. From "F" to "A" on the NY regents science exams: An overview of the aristo project. *arXiv preprint arXiv:1909.01958*.

Clark, P., Etzioni, O., Khot, T., Sabharwal, A., Tafjord, O., Turney, P., and Khashabi, D. 2016. Combining retrieval, statistics, and inference to answer elementary science questions. *Proceedings of AAAI*.

Dagan, I., Roth, D., Sammons, M., and Zanzoto, F. M. 2013. Recognizing textual entailment: Models and applications. *Synthesis Lectures on Human Language Technologies*, *6*(4), 1–220.

Devlin, J., Chang, M.-W., Lee, K., and Toutanova, K. 2018. Bert: Pre-training of deep bidirectional transformers for language understanding. *Proceedings of NAACL*.

Ferreira, W. and Vlachos, A. 2016. Emergent: A novel data-set for stance classification. Pages 1163–1168 of: *Proceedings of NAACL*.

Fletcher, R., and Nielsen, R. . 2018. Automated serendipity: The effect of using search engines on news repertoire balance and diversity. *Digital Journalism*, **6**(8), 976–989.

Gardner, M., Artzi, Y., Basmov, V., Berant, J., Bogin, B., Chen, S., Dasigi, P., Dua, D., Elazar, Y., Gottumukkala, A., Gupta, N., Hajishirzi, H., Ilharco, G., Khashabi, D., Lin, K., Liu, J., Liu, N. F., Mulcaire, P., Ning, Q., Singh, S., Smith, N. A., Subramanian, S., Tsarfaty, R., Wallace, E., Zhang, A., and Zhou, B. 2020. Evaluating models' local decision boundaries via contrast sets. Pages 1307–1323 of: *Proceedings of the 2020 Conference on Empirical Methods in Natural Language Processing: Findings*.

Hasan, K. S. and Ng, V. 2014. Why are you taking this stance? Identifying and classifying reasons in ideological debates. *Proceedings of EMNLP*.

Heckman, J. 1979. Sample selection bias as a specification error. *Econometrica: Journal of the Econometric Society*, **47**(1), 153–161.

Hua, X. and Wang, L. 2017. Understanding and detecting supporting arguments of diverse types. Pages 203–208 of: *Proceedings of ACL*.

Karimi, H., Roy, P., Saba-Sadiya, S., and Tang, J. 2018. Multi-source multi-class fake news detection. Pages 1546–1557 of: *Proceedings of COLING*.

Liu, Y., Ott, M., Goyal, N., Du, J., Joshi, M., Chen, D., Levy, O., Lewis, M., Zettlemoyer, L., and Stoyanov, V. 2019. Roberta: A robustly optimized bert pretraining approach. *arXiv preprint arXiv:1907.11692*.

Mitra, T. and Gilbert, E. 2015. CREDBANK: A large-scale social media corpus with associated credibility annotations. Pages 258–267 of: *ICWSM*. Proceedings of AAAI.

Mohammad, S., Kiritchenko, S., Sobhani, P., Zhu, X., and Cherry, C. 2016. SemEval-2016 Task 6: Detecting stance in tweets. Pages 31–41 of: *Proceedings of SemEval*.

Palau, R. M., and Moens, M.-F. 2009. Argumentation mining: The detection, classification and structure of arguments in text. Pages 98–107 of: *Proceedings of IAAIL*. ACM.

Park, J. and Cardie, C. 2014. Identifying appropriate support for propositions in online user comments. Pages 29–38 of: *Proceedings of the First Workshop on Argumentation Mining*.

Pasternack, J. and Roth, D. 2010. Knowing what to believe (when you already know something). Pages 877–885 of: *Proceedings of the 23rd International Conference on Computational Linguistics (COLING '10)*. USA: Association for Computational Linguistics.

Pasternack, J. and Roth, D. 2014. Judging the veracity of claims and reliability of sources with fact-finders. Pages 39–72 of: *Computational Trust Models and Machine Learning*. Chapman and Hall/CRC.

Rinott, R., Dankin, L., Perez, C. A., Khapra, M., Aharoni, E., and Slonim, N. 2015. Show me your evidence – an automatic method for context dependent evidence detection. Pages 440–450 of: *Proceedings of EMNLP*.

Samadi, M., Talukdar, P., Veloso, M., and Blum, M. 2016. ClaimEval: Integrated and flexible framework for claim evaluation using credibility of sources. Pages 222–228 of: *Proceedings of AAAI*.

Sobhani, P., Inkpen, D., and Zhu, X. 2017. A dataset for multi-target stance detection. Pages 551–557 of: *Proceedings of EACL, Vol. 2*.

Stab, C., Daxenberger, J., Stahlhut, C., Miller, T., Schiller, B., Tauchmann, C., Eger, S., and Gurevych, I. 2018. ArgumenText: Searching for arguments in heterogeneous sources. Pages 21–25 of: *Proceedings of NAACL (Demonstrations)*.

Steiner, M., Magin, M., Stark, B., and Geiß, S. 2020. Seek and you shall find? A content analysis on the diversity of five search engines' results on political queries. *Information, Communication & Society*, 1–25.

Vlachos, A. and Riedel, S. 2014. Fact checking: Task definition and dataset construction. Pages 18–22 of: *Proceedings of the Workshop on Language Technologies and Computational Social Science*.

Vydiswaran, V. G. V., Zhai, C., Roth, D., and Pirolli, P. 2014. Overcoming bias to learn about controversial topics. *Journal of the American Society for Information Science and Technology*, **66**(8), 1655–1672.

Wachsmuth, H., Potthast, M., Al Khatib, K., Ajjour, Y., Puschmann, J., Qu, J., Dorsch, J., Morari, V., Bevendorff, J., and Stein, B. 2017b. Building an argument search engine for the Web. *Workshop on Argument Mining*.

Wang, W. Y. 2017. "Liar, liar pants on fire": A new benchmark dataset for fake news detection. Pages 422–426 of: *Proceedings of ACL, Vol. 2*.

Yin, W. and Roth, D. 2018. TwoWingOS: A two-wing optimization strategy for evidential claim verification. Pages 105–114 of: *Proceedings of EMNLP*. Brussels, Belgium: Association for Computational Linguistics.

16

Formal Representation and Extraction of Perspectives

Aldo Gangemi and Valentina Presutti

16.1 Viewpoints and Perspectives: A Landscape

Humans tend to process multimodal input within stories as a tool to make sense of complex environmental and internal information, and their interaction. Narratological and social cognition research (Boyd, 2009; Dancygier et al., 2016; Enfield and Levinson, 2006) has relevant results on the role played by narratives, and viewpoint occupies a central place (Verhagen, 2016):

> In processing narrative discourse, listeners/readers construct conceptualizations of the ways […] different viewpoints are connected into a meaningful fabric, and moreover connect it to their own point of view, thus adding a further dimension of meaning.

The scale of the Web has changed the way we do research on cognitive phenomena, and requires that we model perspectival discourse so that we can use computation to study it. In computer science, and especially semantic technologies (including, e.g., knowledge representation, knowledge extraction, and natural language processing), we aim at understanding human conceptualization as the optimal way to understand system requirements and the domain in which they apply. Since human viewpoint conceptualization is so widespread, we need to treat viewpoints as first-order entities in our semantic representation of the world. The work so far has scratched the surface of global phenomena such as bias in news (Ma and Yoshikawa, 2009) and viewpoints in opinionated text (Paul et al., 2010). There have been multiple attempts to provide a context semantics to knowledge representation (Hayes, 2012; Klarman and Gutierrez-Basulto, 2016), but the most successful has been in adding syntactic facilities to boost knowledge graphs (Noy et al., 2019) with annotations of statements, whose semantics is, however, entirely left to the user. The recent introduction of RDF* (Olaf Hartig, 2021) helps in reducing the required syntactic overload for linked data.

Since viewpoints/perspectives occupy such a central role in cognition, they escape a simple way of identifying a closed set of semiotic features. Orderly encoding of viewpoints exists in Direct and Indirect Discourse structures (as in the sentences *I am lost, he said he was fainting*), where the provenance of a fact or judgment is explicit. But when multiple viewpoints are reported, things can get complicated. Furthermore, as described in pioneering work by Jakobson (Jakobson, 1957), viewpoints are often contained within Free Indirect Discourse (neither Direct nor Indirect), which "mixes" the provenance of reported facts (so inducing an epistemological quagmire), as in the sentence *She was lost in thought*, where we can only unsafely assume that the author of the text is the holder of the perspective on *she*'s state of mind.

Beyond natural language, perspectives are the primary tool for visual narratives in photography and video, as with the Kuleshov Effect (Mobbs et al., 2006), which also evidences multimodal interferences in originating interpretive perspectives. Perspectives are also present in data, typically in the form of provenance declarations,[1] or when alternative datasets or models on a same topic are integrated.

Far from being only a matter of provenance, perspectives arise from multiple dimensions. In natural language, a plethora of subtle lexical, modal, mood, deictic, as well as discourse markers and cues require attention in order to interpret viewpoints. In data, even when provenance is given, different conceptual schemas, *views*, or queries on data can generate different ontologies (world representations), and, for example, different infographics may eventually lead to biased or distorted interpretations by readers (Krauss, 2012).

A central issue in perspective research is about what makes a perspective emerge, that is, *perspective generators*. As a quick list of cases in perspectival semantics, we provide the following sentences, images, videos, and complex communication phenomena. Far from being complete, we classify them by the kind of *perspective generator* they feature:

- discourse markers: *a cat is on the mat* vs. *that cat is again on the mat*
- lexical frames: *your building is on the hill* vs. *your building hangs over the city*
- attitudes: *I hate speaking in public, I am happy he has been condemned*
- patternicity: use of commonplaces, idioms, sayings, to interpret a situation, for example, *Queen is a fake rock band, The Procrustean bed of crisis: the momentum of a conflict and its analytical fixation inexorably escape each other*

[1] `www.w3.org/TR/prov-dm/#dfn-provenance`.

- culturally sensible framing: biases as emotionally laden framing used against certain people or topics, for example, the *migrants are invaders* framing in political discourse from US, Hungary, Italy in 2018
- alternative framing and storytelling: political talk, for example, *conservatives claim that we need tax relief* vs. *democrats to claim that taxes are investments*; see also the neurocognitive reconstruction of framing and storytelling cases in Lakoff (2008)
- explicit multimodal framing on a neutral fact: the Kuleshov Effect (Mobbs et al., 2006). Kuleshov noticed the multiple meanings emerging out of the contrast and ordering of images in the films of Russian formalists in the early twentieth century
- explicit alternative narratives: Akira Kurosawa's *Rashomon* movie, Escher's *stairways* graphics, the CAMEO ontology (Event Data Project Department, 2012) in the political, diplomatic, and military mediation and conflict domain, alternative reconstructions of facts and causality in science, law, economics, and so on
- silencing: Cancel (or Call-Out) culture, for example, completely ignoring or suppressing the communication/expertise of those who infringe a taboo in social media (or more traditionally, in social circles), for example, attacking a beloved or influential person, defending criminals, committing a hateful crime, behaving against the currently accepted practices, and so on

A theory for public perspectivization practices that have huge effects on societies is emerging from work in social cognition and cognitive neuroscience, with Lakoff's work (Lakoff, 2008) being an interesting attempt toward a unified account: framing works in order to provide individual counterparts to reality (see also Gangemi (2020)) and becomes activated in biological mechanisms such as neural binding, emotional paths, somatic markers, and mirror neurons. Neural binding allows us to "connect the pieces" that come from perception, recall, abstraction, and imagination. Neural binding works according to emotional paths (dopaminergic, noradrenergic), and is linked to narratives and frames. Mirror neurons activate the same circuitry for actual perception, recall of perception, abstraction of perception, and imagination of new perceptions.

We are far from an explanatory and predictive theory that spans from semiotics to neural activation, social cognition, and public behavior; however, the bits and pieces of biological and phenomenological evidence are promising, and make us feel entitled to propose a formal contribution to jointly study perspectivization phenomena.

16.2 What is a Perspective?

This initial landscape analysis of perspective/viewpoint research shows how challenging it is to formally represent perspectives, as well as to single out one or more sets of features that can be used for extracting perspectives. A general definition of viewpoint in language (Dancygier et al., 2016) is that it is "present when an expression represents a person's judgement or when that person is responsible for the expression." This definition hints at viewpoint holders as provenance points for either an expression or a judgment. But a judgment is itself conveyed in an expression. We clearly need a more precise definition before proposing a formal/computational treatment of perspectives.

Perspective is typically understood as an attitude, as in the Oxford Dictionary definition:

a particular attitude towards or way of regarding something; a point of view.

Its etymology goes back to *perspicere*, Latin for "looking through," which links the attitude notion back to a metaphoric account of understanding in terms of perceiving, as exemplified, for example, in the MetaNet repository of English metaphors: `Understanding is Perceiving`. However, the notion of "through" seems to imply an effort to disentangle something in order to give it a view.

The two notions of analyzing and having an attitude toward something can be made formal by using two rhetorical devices. Firstly, we need to detach ourselves from the thing analyzed (*having an attitude*). Then, we need to analyze the thing and tell a story that puts some part of the analyzed thing under the focus of our scrutiny, apparently revealing a less known or obscure aspect, or creating a previously blurred relation, asserting a judgment, or simply creating a spotlight on one of the analyzed parts or aspects.

We call this complex rhetorical device *cognitive perspectivization*. In fact, differently from the act of taking a perspective (or viewpoint, standpoint, vantage point, and so on, which all seem metaphorical parts of the `Perceiving` frame) on a physical object or scene, in cognitive perspectivization we always assume a situation as constructed (perceived, judged, negated, inferred, etc.) by someone, typically in contrast to some other perspective, for example, as part of an argumentation.

This contrastive nature of perspectivization seems coherent with its storytelling nature. Even in rigorous domains such as medicine or jurisprudence, perspectivization is a major reasoning tool and emerges as soon as two descriptions of a same case are taken into account. As argued in Gangemi et al. (2001,

2004, 2005), the expertise used to evaluate alternative diagnostic hypotheses on a patient's situation, or to assess conflicting norms to be applied on the same legal state of affairs, requires us to reason at both intensional (conceptual) and extensional (factual) levels, albeit such levels can be recursively expanded, as with meta-norms (on top of alternative norms), epistemological principles (on top of diagnostic hypotheses), and so on.

Perspectives can then be the target of epistemological practices introduced as regulatory and/or coordination measures: scientific methods, objectivity practices, fact-checking, control of bias, sentiment, emotion, etc.

As with the contrasting sentences: *family is the place of love* vs. *love is the place of family*, perspectivization happens when at least two views could be selected from a ground situation (e.g., *location* + *love* + *family*). This also means that we need to postulate the existence of at least two perspectives, either implicit or explicit, in order for one of them to be considered (e.g., the family-oriented one in the example). In other words, perspectivization always entails comparative reasoning. In everyday life, however, this often goes unnoticed. Political spin, for example, exploits rhetorical means (as in the location+love+family example) in order to make one perspective preferred in the understanding of an audience.

In order to implement perspectivization, a situation should be known, or simply expected to be known (this is the *background*), although only some *aspect* (or *cut* in cinematics language) of the background is shared, and appropriate control constructs (e.g., the location roles in the example sentence) are introduced as a *lens* to observe that aspect. The control constructs reflect the *attitude* of the *conceptualizer*, that is, the perspective holder, as well as its *values*, norms, goals, and so on.

This implementation can be preliminarily represented as a conceptual frame (Fillmore, 1982) or knowledge pattern (Clark et al., 2000; Gangemi and Presutti, 2010). A `Perspective` frame includes roles for *eventualities*, their `background`, a `cut` on them, appropriate `lenses`, possible `attitudes`, and of course a `conceptualizer` that holds the perspective. We may imagine though that additional roles may be active: `recipient`, `time`, `provenance`, `trustworthiness`, `community support`, and so on. We can also imagine relations between perspectives, as well as between perspectives and values, ethical principles, norms, political ideology, and so on.

It is remarkable that this definition includes in practice most of our discourse productions, even in the presence of elliptical sentences such as "*wow!*" (requiring substantial context to reconstruct the perspectival situation), or with *In* (Lu, 2017) constructions such as *I have him in the car* or *I have him in the loop*, where a Langacker's Perspective Taking mechanism (Langacker, 1987) is exemplified: in order to have someone in some place, it requires a vantage point by the perspective holder, who shares the same (physical, social, or abstract) space with the situation addressed by the perspective.

In fact, in cognitive linguistics perspectivization is assumed as a comprehensive process of meaning production (Verhagen, 2007):

> a viewpoint is regarded as the cognitive mechanism of construal of an object in discourse from the point of view of the speaker (conceptualiser). The act of construal is performed by the speaker to influence the cognitive state of the listener.

16.3 Representing and Extracting Perspectives

Based on the conceptual frame that we have sketched for a broad notion of perspective, we can start suggesting a formal treatment.

Firstly, we assume a general knowledge representation (KR) style that is widely used in academia, industry, and government, that is, *knowledge graphs* (Noy et al., 2019). We make use in particular of RDF graphs with OWL2 schemas (Motik et al., 2009), but they can be straightforwardly converted into graph databases, conceptual models, or other KR languages. In the text, we use first-order logic for general accessibility.

Secondly, we assume a knowledge pattern framework such that, given the *background knowledge* about an *eventuality*,[2] *cognitive perspectivization* redescribes that eventuality by providing multiple lenses in the form of *constructions*,[3] narratives, stories, and *principles*[4] to tailor a *cut* within the background eventuality, revealing the *attitude* held by the *conceptualizer* toward the cut, and, by extension, to the overall eventuality, typically in contrast to alternative cuts held by others.

For example, assuming the previous *localization+family+love* example as an eventuality with its background knowledge of frames such as *being a couple*, *mutual love*, and *marriage*, a typical conservative politician would make a cut using a localization metaphor as a lens that puts love situations in the "family as place." On the contrary, a typical progressive perspectivization would rather cut it with a same lens, but inverting it, that is, putting family situations in the "love as place." Of course, this cut is supported by different value lenses: the conservative one depending on the *tradition* value lens that favors marriage, the progressive on a *self-direction* value lens that favors tolerance to alternative

2 An eventuality is a catch-all class including events, event types, factoids, and situations, cf. the DOLCE-Zero ontology at `www.ontologydesignpatterns.org/ont/d0.owl`.

3 A construction may be a linguistic, visual, cinematic, gestural, or even multimodal pattern.

4 Principles can be individual, social, and cultural values, norms, social practices, stereotypes, biases, and so on.

institutions to marriage since it prioritizes a more general *openness to change* value over the *conservation* value.[5]

This assumption is formalized as a `CognitivePerspectivization` (`CP`) *n-ary* relation holding between a conceptualizer producing a cut of an eventuality on its background knowledge, by using one or more lenses. Figure 16.1 shows a Graffoo (Falco et al., 2014) diagram that summarizes the OWL2 (Motik et al., 2009) axioms from the CP ontology.

Due to the polymorphism of the `CP` relation (other arguments can be considered such as time, location, provenance, etc.), it is specified here as a multigrade relation (Gangemi, 2020; Oliver and Smiley, 2004), represented as a *SuperDuper* knowledge pattern (Gangemi and Presutti, 2016), also known as *Descriptions* and *Situations*. A super-duper pattern allows (a) to reify the relation both as a class of *situations* and an intensional *description* (a.k.a. *frame*), (b) to reify its instance relationships as individual situations, and (c) to project all its arguments as both binary relations, and intensional concepts. This solution provides a first-order representation for both the intension of a perspective (its roles, types, relations to other perspectives, etc.) and its extension (its situations). In OWL2, we use *punning* to grant a double interpretation to the same constant used as an individual (a CP frame), and as a class of situations (the CP frame occurrences).

We provide here a first-order axiomatization of the `CP` knowledge pattern. We then provide a frame-based treatment of its intensional part. Each frame from linguistics literature is also represented as a knowledge pattern, following the approach described in Nuzzolese et al. (2011), and implemented in the large Framester knowledge graph (Gangemi et al., 2016). We finally exemplify it with a perspectivization case.

Axiomatization of CP. The `CP` relation (Axioms 1.1 and 1.2) can be formally described as follows: a situation s, in which a conceptualizer c has an attitude a toward a cut ct of an eventuality e, jointly with its background knowledge b, shot by means of a lens l, possibly in contrast[6] to some other (either explicit or implicit) cut ct_1 that addresses at least part of the same background b (e.g., a same set of facts with causal relations among them), but possibly with a different lens l_1. Axiom 1.1 is the extensional part of CP, while axiom 1.2 provides the intensional types. Axioms 1.3 and 1.4 provide the semantic types for the ranges of semantic roles.

[5] We are using here the circumplex model of Schwartz' (2012) as a reference for those values.

[6] A contrast is not necessarily a source of conflict: contrast relations are a matter of research, which should be investigated empirically, for example, from distributional evidence and psychological testing.

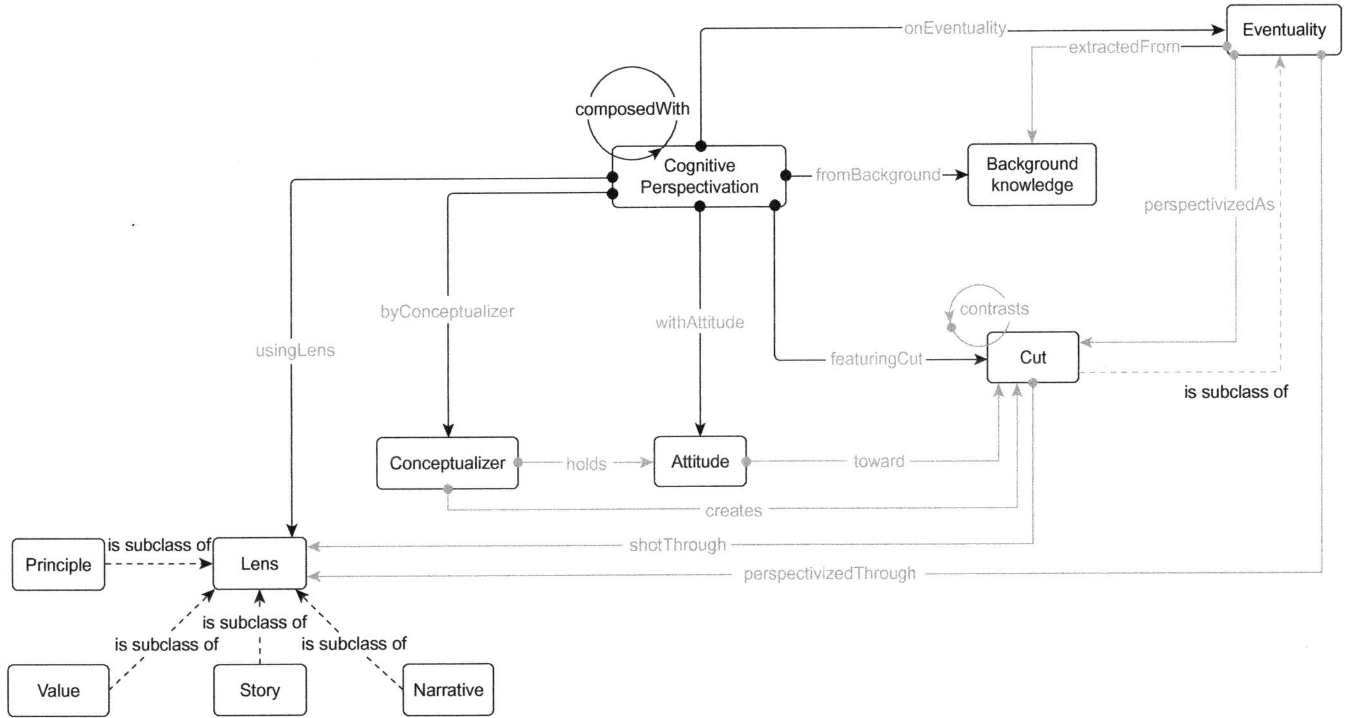

Figure 16.1 A diagrammatic summary of the Cognitive Perspectivization knowledge pattern. Bold arrows denote the main arguments of the CP relation; the others are projection or subsumption relations

$$\forall(s)\,\mathtt{CP}(s) \rightarrow \mathtt{occurrenceof}(s,CP)\wedge \\ \exists(c,e,b,l,a,ct)\,\mathtt{conceptualizer}(s,c)\wedge\mathtt{eventuality}(s,e)\wedge \\ \mathtt{background}(s,b)\wedge\mathtt{lens}(\mathtt{s},\mathtt{l})\wedge\mathtt{attitude}(s,a)\wedge\mathtt{cut}(s,ct) \tag{16.1}$$

$$\mathtt{CP}(s) \rightarrow \mathtt{Frame}(\mathtt{CP})\wedge \\ \mathtt{hasProjection}(\mathtt{CP},r)\wedge\mathtt{SemanticRole}(r), \\ r\in\{\mathtt{lens},\mathtt{cut},\mathtt{eventuality}, \\ \mathtt{conceptualizer},\mathtt{background},\mathtt{attitude}\} \tag{16.2}$$

$$(r(s,y)\wedge r\neq\mathtt{conceptualizer})\rightarrow\mathtt{Situation}(y) \tag{16.3}$$

$$\mathtt{conceptualizer}(s,y)\rightarrow\mathtt{Agent}(y) \tag{16.4}$$

The argument projections of CP shown in Figure 16.1 can be axiomatized as in Axioms 1.5–1.11 (assuming previous axioms). A contrast relation between cuts that are not related by subsumption is defined in Axiom 1.12.

$$\forall(s,e,b)\,(\mathtt{eventuality}(s,e)\wedge\mathtt{background}(s,b))\rightarrow \\ \mathtt{extractedFrom}(e,b) \tag{16.5}$$

$$\forall(s,e,ct)\,(\mathtt{eventuality}(s,e)\wedge\mathtt{cut}(s,ct))\rightarrow \\ \mathtt{perspectivizedAs}(e,ct) \tag{16.6}$$

$$\forall(s,e,l)\,(\mathtt{eventuality}(s,e)\wedge\mathtt{lens}(\mathtt{s},\mathtt{l}))\rightarrow \\ \mathtt{perspectivizedThrough}(e,l) \tag{16.7}$$

$$\forall(sl,ct)\,(\mathtt{lens}(\mathtt{s},\mathtt{l})\wedge\mathtt{cut}(s,ct))\rightarrow\mathtt{shotThrough}(ct,l) \tag{16.8}$$

$$\forall(s,c,a)\,(\mathtt{conceptualizer}(s,c)\wedge\mathtt{attitude}(s,a))\rightarrow \\ \mathtt{holds}(c,a) \tag{16.9}$$

$$\forall(s,a,ct)\,(\mathtt{attitude}(s,a)\wedge\mathtt{cut}(s,ct))\rightarrow\mathtt{toward}(a,ct) \tag{16.10}$$

$$\forall(s,c,a,ct)\,(\mathtt{holds}(c,a)\wedge\mathtt{toward}(a,ct))\rightarrow\mathtt{creates}(c,ct) \tag{16.11}$$

$$\forall(ct,ct_1)\,(\mathtt{shotThrough}(ct,l)\wedge\mathtt{shotThrough}(ct_1,l_1)\wedge \\ \neg((\mathtt{subsumes}(l,l_1)\cup\mathtt{subsumes}(l_1,l))\leftrightarrow \\ \mathtt{contrasts}(ct,ct_1) \tag{16.12}$$

$$\forall(e,l)\mathtt{perspectivizedThrough}(e,l)\leftrightarrow e\otimes l=ct \tag{16.13}$$

A cut *ct* can also be viewed as an occurrence of a composition of an eventuality with a lens: the cinematic metaphor here well describes the Kuleshov conceptual effect at work in perspectivization. Axiom 1.13 uses a compositional operator $\otimes$ between situations, as a consequence of the cut notion as a perspectivization result on an eventuality, shot through a lens. The

CP frame has been implemented in OWL2, and aligned to the Framester schema.[7]

Aligning CP to Known Frames and the Emergence of Compositionality. Most newly introduced frames/KP can be aligned to existing ones. On one hand, the FrameNet lexical resource (Baker et al., 1998) has been the first to (informally) systematize known frames, followed by other linguistic resources, and eventually addressing also *frame matching* problems as with metaphors, for example, with MetaNet (David et al., 2014). On the other hand, formal ontology, ontology engineering, and the Semantic Web, even without explicitly adopting Fillmore's frame semantics, have eventually formalized a huge amount of knowledge patterns (see, e.g., the Ontology Design Patterns Repository,[8] as well as examples of KPs that underlie foundational ontologies such as DOLCE (Presutti and Gangemi, 2016)), which also constitute precedents to any newly proposed frame. A formal integration of the different sources for frames, KPs, lexical resources, and so on is provided by resources such as Framester (Gangemi et al., 2016),[9] and Premon (Rospocher et al., 2019). Framester is used here because its semantics fully integrates data semantics and lexical semantics.

In the case of Cognitive Perspectivization, we can look for the FrameNet frames, to which we can align the CP frame, and then we may wonder whether that alignment can help us in operationalizing the CP by assisting semantic technologists in using, for example, joint data and text information to reason on perspectivization.

Firstly, frame alignment provides new potential axioms that could enrich, or induce a refactoring of, a proposed frame. Frame alignment works by detecting the closest frames in FrameNet which match the roles from CP (see below), and aligning them.

Secondly, frame composition can help us in operationalizing CP. We start with a minimal theory of frame composition (Gangemi, 2020). Axiom (1.14) introduces a $\otimes$ compositional binary operator stating that any two frames f and g are composed if and only if at least one role r from f and one role s from g are composed. When two roles are composed, their filler x in a frame composition occurrence (an eventuality $e^l \neq e$) is the same (Axiom 1.15). A composition creates a new frame/role (Axiom 1.16).

[7] www.ontologydesignpatterns.org/ont/persp/perspectivization.owl.
[8] www.ontologydesignpatterns.org.
[9] The Framester knowledge graph can be queried from http://etna.istc.cnr.it/framester2/sparql, while its ontology schema is available at https://w3id.org/framester/schema/.

$$\forall(f,g)(f \otimes g) \leftrightarrow \exists(r,s)(\texttt{hasRole}(f,r) \wedge \texttt{hasRole}(g,s) \wedge (r \otimes s)) \quad (16.14)$$

$$\forall(r,s)(r \otimes s) \leftrightarrow \exists(e)(\texttt{r.F}(e,x) \wedge \texttt{s.G}(e,x)) \quad (16.15)$$

$$\forall(x,y)(x \otimes y)(\texttt{Frame}(x) \wedge \texttt{Frame}(y) \rightarrow \exists(z)(\texttt{Frame}(z)) \cup (\texttt{Role}(x) \wedge \texttt{Role}(y)) \rightarrow \exists(z)(\texttt{Role}(z))) \quad (16.16)$$

For example, when `CP` is aligned to its closest matches from Framester, perspectives seem to depend at least on the following FrameNet frames: `Attributed_information` (for the conceptualization part), `Mental_property` (for the attitude part), `Scrutiny` (for the cut part), and `Differentiation` (for the contrasting/similarity part):

- `Regarding` and `Representing` approximate the lens part of CP. An Entity represents some Phenomenon through its existence and/or defining characteristics. Entity corresponds to the `CP` cut, Phenomenon to the eventuality.
- `AttributedInformation` has the following given roles: Proposition, Speaker, Text: a Proposition is attributed to a Speaker or directly to a Text (as a provenance attribution). Speaker corresponds to the `CP` conceptualizer, the Proposition to the lens, and the Text to the background.
- `Mental_property` has the following given roles: Judge, Behavior, Protagonist, Degree, Manner, Practice, and Domain; the core ones that are at stake in CP are Judge, Manner, and Behavior. Judge corresponds to the conceptualizer, Manner to the attitude, and Behavior to the cut.
- `Scrutiny` has the following given roles: Cognizer, Phenomenon, Purpose, Time, Medium, Means, Degree, Direction, Ground, Manner, Instrument, with Cognizer, Manner, and Phenomenon corresponding to the `CP` conceptualizer, attitude and eventuality respectively. `Differentiation` has the following given roles: Cognizer, Phenomena, Quality, Degree, Means, Manner, Circumstances, Depictive, with Cognizer, Phenomenon1, Phenomenon2, Circumstances corresponding to the conceptualizer, cut, and background respectively.

We may briefly exemplify an application of `CP` alignment with Axiom 1.17, which associates `CP` with the `Attributed_information` frame by composing some of the roles introduced in Axiom 1.1.

$$\forall(s,p,a,ct,b)\,\texttt{CP}(s,p,a,ct,b) \rightarrow \texttt{conceptualizer.CP}(s,p) \leftrightarrow \texttt{speaker.AI}(s,p) \wedge \quad (16.17)$$

$$\texttt{cut.CP}(s,ct) \leftrightarrow \texttt{proposition.AI}(s,ct) \wedge$$
$$\texttt{background.CP}(s,b) \leftrightarrow \texttt{text.AI}(s,b)$$

Frame Evocation with Cognitive Perspectivisation. As described, frame alignment and composition techniques can now be applied for:

1. extracting frame-based knowledge graphs from text with FRED (Gangemi et al., 2017) machine reader, which automatically parses, logically shapes, and links the knowledge expressed by the sentence to public knowledge graphs;
2. anchoring the extracted graph to perspectival knowledge through the large frame-based Framester's knowledge graph.

Framester frames and roles are automatically associated with FRED knowledge graphs via the FRED2Framester API,[10] and when a perspectivization hint is detected in the graph, the `CP` frame is used to overload the knowledge graph with perspectival knowledge. We call that overload *super-dupering*, cf. Section 16.3.

We exemplify this operationalization with respect to a classic perspectivization example from Section 16.2, used by spin doctors to explain storytelling techniques: *Conservatives claim that we need relief from taxes* vs. *Democrats claim that taxes are investments for us*.

We apply the approach as sketched at the beginning of this section by firstly parsing the sentences with a frame-based approach, and extracting two knowledge graphs KG_c and KG_d. Then we apply the double intensional/extensional modeling to the extracted frames, and align them to the *CP* frame. Finally, we analyze the compositionality resulting from aligning the superdupered knowledge graphs to public semantic resources.

Frame-Based Knowledge Extraction. We pass the two sentences to the FRED machine reader API,[11] obtaining two knowledge graphs. FRED graphs contain RDF triples expressing:

- the concepts expressed in the sentences: Claim, Need, Relief, Tax, Investment

[10] `http://wit.istc.cnr.it/stlab-tools/fred_api/`.

[11] A Web application for testing is also available at `http://wit.istc.cnr.it/stlab-tools/fred/`.

- the anonymous referents of those concepts (claims, needs, reliefs, investments, taxes)
- individual entities and facts: Conservatives, Democrats, anonymous persons (we, us) denoted in the sentences
- the roles (agent, experiencer, etc.) evoked by the words in the syntactically dominant frames (Claim, Need)
- the automatically linked entities and concepts from public repositories, which can be used to extend the graph automatically with public background knowledge (e.g., individuals, concepts, and frames related to Claim, Need, Tax, Investment, Conservatism, etc.)

From the first sentence, we extract KG_c, expressible in first-order logic as the existential Axiom 1.18. From the second sentence, we extract KG_d, expressed in axiom 1.19:

$$\begin{aligned} \exists(x,c,p,n,h,r,t)(\texttt{Conservative}(x) \wedge \texttt{Claim}(c) \wedge \\ \texttt{Person}(p) \wedge \texttt{Need}(n) \wedge \texttt{Have}(h) \wedge \texttt{Relief}(r) \wedge \texttt{Tax}(t) \wedge \\ \texttt{agent.Claim}(c,x) \wedge \texttt{theme.Claim}(c,n) \wedge \\ \texttt{experiencer.Need}(n,p) \wedge \texttt{theme.Need}(n,h) \wedge \\ \texttt{agent.Have}(h,p) \wedge \texttt{patient.Have}(h,r) \wedge \\ \texttt{from.Need}(n,t) \end{aligned} \tag{16.18}$$

$$\begin{aligned} \exists(x,c,p,s,t)(\texttt{Democrat}(x) \wedge \texttt{Claim}(c) \wedge \\ \texttt{Person}(p) \wedge \texttt{Situation}(s) \wedge \texttt{Tax}(t) \wedge \\ \texttt{agent.Claim}(c,x) \wedge \texttt{theme.Claim}(c,s) \wedge \\ \texttt{involves}(s,t) \wedge \texttt{involves}(s,p) \wedge \\ \texttt{Investment}(t) \wedge \texttt{for}(t,p) \end{aligned} \tag{16.19}$$

For an intuitive rendering of the semantic subgraphs produced by FRED,[12] cf. Figures 16.2 and 16.3, which also include the entity linking axioms not shown in the FOL axioms.

SuperDupering Knowledge Graphs with the Perspectivization Frame. In the running example, the task of extracting perspectives is quite straightforward because many arguments of CP can be filled with values from the extracted graphs. In other words, perspectivization is explicit in the sentences, for example, the *claim* frame points directly at a perspectivization. In more complex

[12] FRED produces two linked subgraphs, the first annotates textual fragments with semantic and syntactic knowledge, the second (the semantic subgraph) is a frame-based knowledge graph including both a schema and the data extracted from the sentence, aligned to public resources (WordNet, VerbNet, FrameNet, schema.org, DBpedia, etc.).

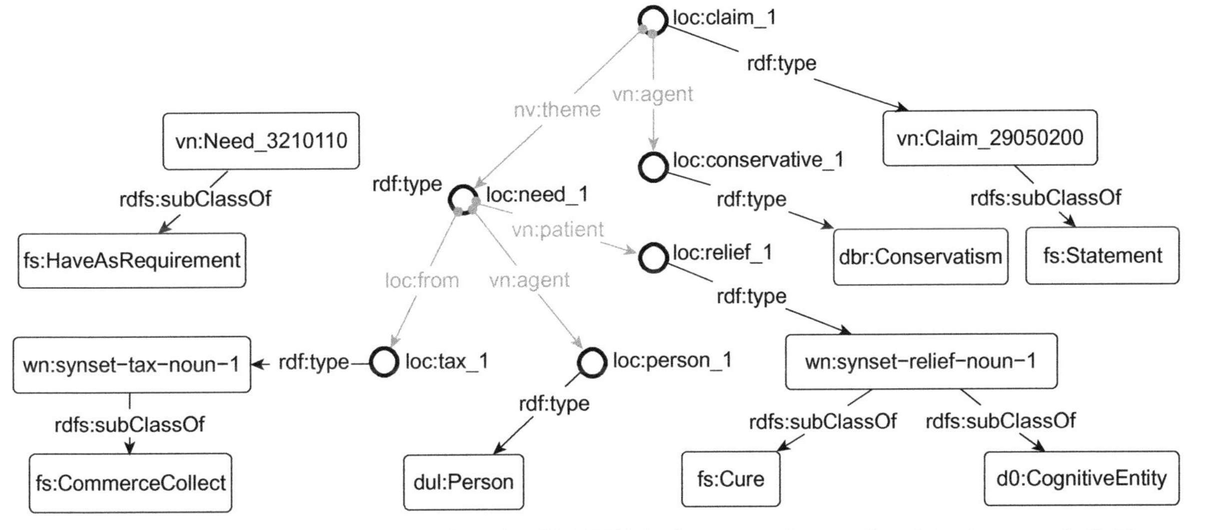

Figure 16.2 A diagram showing the semantic subgraph produced by FRED for the sentence *Conservatives claim that we need relief from taxes*. Circles denote classes; diamonds denote individuals. Prefixes: loc: for local resources, vn: for VerbNet senses, wn: for WordNet synsets, dbr: for DBpedia entities, dul: for DOLCE, d0: for DOLCE-Zero

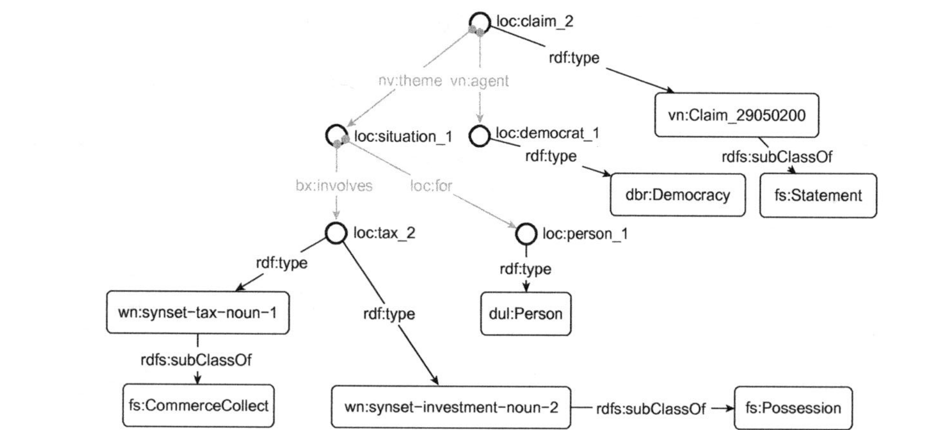

Figure 16.3 A diagram showing the semantic subgraph produced by FRED for the sentence *Democrats claim that taxes are investments for us*

cases, perspectivization hints might be subtler. Axioms 1.20 through 1.24 show how the SuperDupering operates on CP occurrences:

$$\begin{aligned}KG_c \uparrow \exists(e_1,w)(\mathtt{CP}(e_1) \wedge \mathtt{TaxKG}(w)\wedge \\ \mathtt{conceptualizer}(e_1,\mathtt{Conservative})\wedge \\ \mathtt{background}(e_1,w) \wedge \mathtt{attitude}(e_1,\mathtt{Need})\wedge \\ \mathtt{lens}(e_1,\mathtt{Relief}) \wedge \mathtt{eventuality}(e_2,\{\mathtt{Tax}\})\end{aligned} \tag{16.20}$$

$$\begin{aligned}\mathtt{background}(e_1,w) \wedge \mathtt{attitude}(e_1,\mathtt{Need})\wedge \\ \mathtt{lens}(e_1,\mathtt{Relief}) \wedge \mathtt{eventuality}(e_2,\{\mathtt{Tax}\} \rightarrow \\ \mathtt{cut}(e_2,\{\mathtt{Tax} \otimes \mathtt{Relief}\})\end{aligned} \tag{16.21}$$

$$\begin{aligned}KG_d \uparrow \exists(e_2,w)(\mathtt{CP}(e_2) \wedge \mathtt{TaxKG}(w)\wedge \\ \mathtt{conceptualizer}(e_2,\mathtt{Democrat})\wedge \\ \mathtt{background}(e_2,w) \wedge \mathtt{attitude}(e_2,s)\wedge \\ \mathtt{lens}(e_2,\mathtt{Investment}) \wedge \mathtt{eventuality}(e_2,\{\mathtt{Tax}\})\end{aligned} \tag{16.22}$$

$$\begin{aligned}\mathtt{background}(e_1,w) \wedge \mathtt{attitude}(e_1,\mathtt{s})\wedge \\ \mathtt{lens}(e_1,\mathtt{Investment}) \wedge \mathtt{eventuality}(e_2,\{\mathtt{Tax} \rightarrow \\ \mathtt{cut}(e_2,\{\mathtt{Tax} \otimes \mathtt{Investment}\})\end{aligned} \tag{16.23}$$

$$\begin{aligned}KG_c \otimes KG_d \uparrow \mathtt{contrasts}(\{\mathtt{Tax} \otimes \mathtt{Relief}\}, \\ \{\mathtt{Tax} \otimes \mathtt{Investment}\})\end{aligned} \tag{16.24}$$

Based on the CP frame and its axioms, Axioms 1.20 and 1.22 represent the overdescription of the two FRED graphs with CP: a conceptualizer holds an attitude toward a lens applied to a background eventuality, producing a certain cut that makes a blend emerge out of the role composition ($\mathtt{lens} \otimes \mathtt{cut}$), as shown in axioms 1.21 and 1.23. SuperDupering (denoted as $\uparrow$) operates on each knowledge graph, assigning the CP roles to individuals or concepts from the graph whenever possible, or else introducing new individuals. This happens for:

- the CP-ed situations (e_1 and e_2);
- the `background` w, that is, the TaxKG knowledge graph, automatically extracted from the dbpedia Tax graph thanks to the FRED linking, including, for example, Tax_policy, Tax_revenue, Tax_avoidance, and so on, which are not explicit in the sentences. The construction of the TaxKG is a result of the semiotic drifting (cf. next paragraph);
- the `conceptualizer` role "superdupers" conservatives and democrats;

- the `attitude` role superdupers the claims, that is, *need* for Conservatives, and the asserted situation for Democrats;
- the `lens` role superdupers the `Relief` frame for conservatives, and the `Investment` frame for democrats;
- `cut` roles superdupers the Tax⊗Relief frame for conservatives, and the Tax⊗Investment for democrats;
- finally, Axiom 1.24 shows the composition of the two knowledge graph, which superdupers a contrast between the two different cuts, which are not related explicitly in the Framester knowledge graph;[13]
- in this example, the eventuality role filler coincides with the `background` one, because the claims are about a general value, and no specific time is expressed, which could distinguish tax knowledge in general, from current circumstances.

Semiotic Drifting. At this point, we have got two superdupered graphs KG^s_c and KG^s_d. We have used the alignment functionalities of FRED's to disambiguate the concepts and named individuals, and then we have queried the Framester knowledge graph hub to collect all the frames evoked by the sentences, so generating a richer composition. Axioms 1.20 and 1.22 show a sample of those alignments, which exemplify a method used in knowledge extraction which we call *semiotic drifting*, consisting of dynamically expanding a graph g with relevant knowledge as soon as some part of g gets aligned to other knowledge. Expansion can be guided by requirements, opportunity, or just serendipity, and can be implemented with off-the-shelf graph traversal algorithms. Since the amount of alignments can get really huge after a couple of iterations, we show here just a few.

Conservatives and Democrats are easily matched to public DBpedia or Wikidata entities, while the Claim, Need, Relief, Tax, Investment concepts are aligned to WordNet, VerbNet, FrameNet, and ultimately to Framester core frames: VerbNet's `Claim_29050200` (leading to the `Statement` frame) and `Need_32010110` (leading to the `Needing` frame), WordNet's `synset-relief-noun-1` (metaphorically leading to the `Cure` frame by reusing the Framester extension to MetaNet frame-based metaphors, now represented in the Amnestic Forgery knowledge graph (Gangemi et al., 2018)), DBpedia's `Conservatism` and `Democracy` (leading to the `PeopleAlongPoliticalSpectrum` frame), Tax (leading to the `EarningsAndLosses` frame), and Investment (leading to the `Possession` frame).

[13] I.e., each of them does not subsume, use, etc. the other.

Rule-Based Superdupering The superdupered mapping of extracted entities and concepts to the `CP` roles is currently performed in a rule-based form, for example, the `attitude` role is associated with modal expressions, the `conceptualizer` role with agents of perspectivization frames (e.g., *verba dicendi*), the `background` role with semiotic drifting (see below), the `lens` role is, for example, assigned in contrast to the *eventuality* role, according to role chains of frame composition in discourse. Finally, the `cut` role is differentially established according to the composition of explicit frames.

However, the superdupered mapping of extracted entities and concepts to the `CP` roles needs further research in order to design or learn rules that have enough variability to deal with complex perspective networks. Rhetorical Structure Theory (Ji and Smith, 2017) and Argumentation Mining (Stede and Schneider, 2018) might provide some heuristics, as well as a large-scale exploration starting with, for example, the training of an LSTM to recognize patterns of `CP` roles.

Extraction Strategies: A Generalization. Existing text mining approaches that deal with viewpoints address biases (Ma and Yoshikawa, 2009), opinions (Paul et al., 2010), and so on, but they hardly attack the problem analytically (an exception is Gangemi et al. (2014)); even smaller literature targets multimodal perspectives (a recent work is Zadeh et al. (2017)). Because of the variety of multimodal expressions that can evoke a perspective, we barely need methodological hybridization, which can maximize recall in detecting or discovering perspectivization with inductive techniques.

On the other hand, we are injecting semantics into feature engineering, for example, by using techniques such as graph embeddings over knowledge graphs (Goyal and Ferrara, 2018; Ristoski and Paulheim, 2016). The SuperDuper approach proposed here, jointly with a hybrid semiotic drifting, can provide a healthy feedback loop in both discovering and formally reasoning over features. `CP`-like frames can also be used as know-how repositories for experiments over perspectival thinking.

16.4 Conclusions

In this chapter, we have presented a brief assessment of current linguistic, cognitive, and computational literature on perspectivization, as well as a pragmatic method to extract and represent perspectival knowledge.

The major difficulty in making computational sense of perspectives is that they are (1) widespread in communication; (2) multimodal; (3) only interpretable with respect to implicit, indirect, or nuanced contextual knowledge.

The differences between the many types of perspectivization phenomena seem huge, but, inspired by cognitive linguistics, frame semantics, and pattern-based ontology design, we have proposed a SuperDuper approach to perspective representation, which admits multilayered framing that can be used across multiple types of perspectivization.

Our main intention is to show the feasibility of SuperDuper-ing knowledge representation without adding logical complexity and also retaining the basic expressivity that can be employed in both scalable knowledge engineering methods and feature engineering methods for machine learning.

References

Baker, C. F., Fillmore, C. J., and Lowe, J. B. 1998. The Berkeley FrameNet project. Pages 86–90 of: *Proceedings of the 36th Annual Meeting of the Association for Computational Linguistics and 17th International Conference on Computational Linguistics, Vol. 1*. Montreal, Quebec, Canada: Association for Computational Linguistics.

Boyd, B. 2009. *On the origin of stories: Evolution, cognition and fiction*. Cambridge, MA: The Belknap Press of Harvard University Press.

Clark, P., Thompson, J., and Porter, B. 2000. Knowledge patterns. Pages 591–600 of: Cohn, A. G., Giunchiglia, F., and Selman, B. (eds), *KR2000: Principles of knowledge representation and reasoning*. San Francisco: Morgan Kaufmann.

Dancygier, B., Lu, W., and Verhagen, A. (eds). 2016. *Viewpoint and the fabric of meaning: Form and use of viewpoint tools across languages and modalities*. Berlin: De Gruyter Mouton.

David, O., Dodge, E., Hong, J., Stickles, E., and Sweetser, E. 2014. Building the MetaNet metaphor repository: The natural symbiosis of metaphor analysis and construction grammar. Pages 3–6 of: *8th International Conference on Construction Grammar (ICCG 8)*. Osnabrück: ICLA International Cognitive Linguistics Association.

Enfield, N. J. and Levinson, S. C. (eds). 2006. *Roots of human sociality: Culture, cognition and interaction*. Oxford: Berg.

Event Data Project Department. 2012 (March). CAMEO: Conflict and mediation event observations: Event and actor codebook, version 1.1b3. Technical report Political Science Pennsylvania State University Pond Laboratory University Park, PA.

Falco, R., Gangemi, A., Peroni, S., Shotton, D. M., and Vitali, F. 2014. Modelling OWL ontologies with graffoo. Pages 320–325 of: Presutti, V., Blomqvist, E., Troncy, R., Sack, H., Papadakis, I., and Tordai, A. (eds), *The Semantic Web: ESWC 2014 Satellite Events - Anissaras, Crete, Greece, May 25–29, 2014, Revised Selected Papers*. Lecture Notes in Computer Science, Vol. 8798. Berlin: Springer.

Fillmore, C. 1982. Frame semantics. Pages 111–137 of: *Linguistics in the morning calm*. Seoul: Hanshin.

Gangemi, A. 2020. Closing the loop between knowledge patterns in cognition and the semantic Web. *Semantic Web*, **11**(1), 139–151.

Gangemi, A., Alam, M., Asprino, L., Presutti, V., and Reforgiato Recupero, D. 2016. Framester: A wide coverage linguistic linked data hub. Pages 239–254 of: *Knowledge Engineering and Knowledge Management - 20th International Conference*. EKAW '16. Springer, Berlin: European Knowledge Acquisition Workshop. 10.1007/978-3-319-49004-5_16.

Gangemi, A., Alam, M., and Presutti, V. 2018. Amnestic forgery: An ontology of conceptual metaphors. Pages 159–172 of: *Formal Ontology in Information Systems - Proceedings of the 10th International Conference, FOIS 2018, Cape Town, South Africa*. Amsterdam: IOS Press. 10.3233/978-1-61499-910-2-159.

Goyal, P., and Ferrara, E. 2018. Graph embedding techniques, applications, and performance: A survey. *Knowledge-Based Systems*, **151**, 78–94.

Gangemi, A., Catenacci, C., and Battaglia, M. 2004. Inflammation ontology design pattern: An exercise in building a core biomedical ontology with descriptions and situations. Pages 64–80 of: Pisanelli, D. M. (ed), *Ontologies in medicine*. Amsterdam: IOS Press.

Gangemi, A., Pisanelli, D. M., and Steve, G. 2001. An ontological framework to represent norm dynamics. Winkels, R. (ed), *Proceedings of the 2001 Jurix Conference, Workshop on Legal Ontologies*. Amsterdam, The Netherlands: University of Amsterdam.

Gangemi, A., and Presutti, V. 2010. Towards a pattern science for the semantic Web. *Semantic Web*, **1**(1–2), 61–68. 10.3233/SW-2010-0020.

Gangemi, A., and Presutti, V. 2016. *Multi-layered n-ary patterns*. Pages 105–131 of: *Ontology engineering with ontology design patterns*. IOS Press.

Gangemi, A., Presutti, V., and Recupero, D. R. 2014. Frame-based detection of opinion holders and topics: A model and a tool. *IEEE Computational Intelligence Magazine*, **9**(1), 20–30. 10.1109/MCI.2013.2291688.

Gangemi, A., Presutti, V., Recupero, D. R., Nuzzolese, A. G., Draicchio, F., and Mongiovì, M. 2017. Semantic Web machine reading with FRED. *Semantic Web*, **8**(6), 873–893. 10.3233/SW-160240.

Gangemi, A., Sagri, M. T., and Tiscornia, D. 2005. A constructive framework for legal ontologies. Pages 97–124 of: *Law and the Semantic Web*, Vol. LNCS 3369. Berlin: Springer.

Hartig, O. and Champin, P.-A. 2021. RDF-star and SPARQL-star: Draft community group report 05 March 2021. Technical report. W3C.

Hayes, P. 2012. Situations, contexts, states of affairs, and the limits of formalization. *IEEE International Multi-Disciplinary Conference on Cognitive Methods in Situation Awareness and Decision Support*. New Orleans, Louisiana: IEEE Catalog.

Jakobson, R. 1957. Shifters, verbal categories and the Russian verb. Russian Language Project. Cambridge: Harvard University, Department of Slavic Languages and Literatures.

Ji, Y., and Smith, N. 2017. Neural discourse structure for text categorization. *arXiv preprint arXiv:1702.01829*.

Klarman, S. and Gutierrez-Basulto, V. 2016. Description logics of context. *Journal of Logic and Computation*, **26**(3), 817–854.

Krauss, J. 2012. Infographics: More than words can say. *Learning & Leading with Technology*, **39**(5), 10–14.

Lakoff, G. 2008. *The political mind.* New York: Viking Press.

Langacker, R. 1987. *Foundations of cognitive grammar, Vol. 1.* Redwood: Stanford University Press.

Lu, W.-L. 2017. Perspectivization and contextualization in semantic analysis: A parsimonious polysemy approach to in. *Studia Linguistica Universitatis Iagellonicae Cracoviensis*, **134**(1), 247–264.

Ma, Q., and Yoshikawa, M. 2009. Topic and viewpoint extraction for diversity and bias analysis of news contents. Pages 150–161 of: Li, Q., Feng, L., Pei, J., Wang, S. X., Zhou, X., and Zhu, Q.-M. (eds), *Advances in data and Web management.* Berlin: Springer.

Mobbs, D., Weiskopf, N., Lau, H., Featherstone, E., Dolan, R., and Frith, C. 2006. The Kuleshov effect: The influence of contextual framing on emotional attributions. *Social Cognitive and Affective Neuroscience*, **1**, 95–106.

Motik, B., Parsia, B., and Patel-Schneider, P. F. 2009. OWL 2 Web ontology language: Structural specification and functional-style syntax. *W3C Recommendation*, **27**(65), 159.

Noy, N., Gao, Y., Jain, A., Narayanan, A., Patterson, A., and Taylor, J. 2019. Industry-scale knowledge graphs: Lessons and challenges. *Queue*, **17**(2), 95–106.

Nuzzolese, A. G., Gangemi, A., and Presutti, V. 2011. Gathering lexical linked data and knowledge patterns from FrameNet. Pages 41–48 of: *Proceedings of the 6th International Conference on Knowledge Capture (K-CAP 2011), June 26–29, 2011, Banff, Alberta, Canada.* 10.1145/1999676.1999685.

Oliver, A., and Smiley, T. 2004. Multigrade predicates. *Mind*, **113**(452), 609–681.

Paul, M. J., Zhai, C. X., and Girju, R. 2010. Summarizing contrastive viewpoints in opinionated text. Pages 66–76 of: *Proceedings of the 2010 Conference on Empirical Methods in Natural Language Processing.* Massachusetts: Association for Computational Linguistics.

Presutti, V., and Gangemi, A. 2016. Dolce+D&S ultralite and its main ontology design patterns. Pages 81–103 of: *Ontology Engineering with Ontology Design Patterns - Foundations and Applications.* 10.3233/978-1-61499-676-7-81.

Ristoski, P., and Paulheim, H. 2016. RDF2Vec: RDF graph embeddings for data mining. Pages 498–514 of: Groth P. et al. (eds). The Semantic Web – ISWC 2016. ISWC 2016. Lecture Notes in Computer Science, Vol. 9981. Cham: Springer.

Rospocher, M., Corcoglioniti, F., and Aprosio, A. P. 2019. PreMOn: LODifing linguistic predicate models. *Language Resources and Evaluation*, **53**(3), 499–524.

Schwartz, S. H. 2012. An overview of the Schwartz theory of basic values. *Online Readings in Psychology and Culture*, **2**(1). 10.9707/2307-0919.1116.

Stede, M., and Schneider, J. 2018. Argumentation mining. *Synthesis Lectures on Human Language Technologies*, **11**(2), 1–191.

Verhagen, A. 2007. *Construal and perspectivisation.* Oxford: Oxford University Press. Pages 48–81.

Verhagen, A. 2016. *Introduction: On tools for weaving meaning out of viewpoint threads.* Berlin: De Gruyter Mouton. Pages 1–10.

Zadeh, A., Chen, M., Poria, S., Cambria, E., and Morency, L. P. 2017. *Tensor fusion network for multimodal sentiment analysis.* arXiv preprint arXiv:1707.07250.

17

The User Perspective in Professional Information Search

Suzan Verberne

17.1 Introduction

Information retrieval (IR) is the research field that addresses the development, optimization, and evaluation of search engines and the study of how humans interact with search engines. Traditionally, search systems were investigated in the context of libraries and librarians (Harman et al., 2019), but the main focus of the field since the 1990s has been on Web search engines that serve an immense target audience (e.g., Google).

A large part of IR research is involved in the development of ranking optimization methods – a machine learning – driven task. But the field has never lost sight of the user perspective (Croft, 2019; Ingwersen and Järvelin, 2005). The human is central in the information search process because the user formulates queries, views the retrieved documents, judges their relevance, and decides when to stop searching (Maxwell, 2019). The user's perspective on information is important in the IR process in two different ways: (1) the user formulates a search query based on their own perspective of the task at hand and the required information; and (2) the user assesses the results returned by the search engine for relevance to their information need. Depending on the user's perspective, a document could be relevant or irrelevant to the entered query. Although the user acts from their personal perspective, the search engine only sees the user's interactions: the entered queries and the clicked documents. For example, consider a user entering the query "Rembrandt Leiden" in a Web search engine. The perspective of the user could be a historical one, searching information about Rembrandt van Rijn's youth in Leiden. It could also be a touristic perspective, searching for Rembrandt locations in Leiden that are worth visiting. Or maybe the user's goal is simply to navigate to the Web page of the restaurant Rembrandt in Leiden.

In this chapter, we use the term *perspective* in the context of IR to refer to all user aspects that lead to the formulated query and the assessment of the results: the user's interest, background information, current task context, and information need.

Because of the central role of the user in IR, it feels natural to take the individual user's perspective into account in the development of search engines. Current search engines use a form of personalization in the ranking of the search results, adapting the ranking for a small portion of the search results. This form of personalization could, for example, accommodate for local search, ranking search results from locations close to the user higher than results that are further away (Hannak et al., 2013). But the large majority of the search results in Web search engines are in fact user-independent: each user receives the same results without any personalization. The ranking is strongly effected by the popularity of pages – estimated by the clicks of other users (Joachims et al., 2005). This is effective for common Web search tasks but not for the highly specific search tasks performed in professional contexts.

Professional search is the searching carried out by experts for work purposes (Russell-Rose et al., 2018; Verberne et al., 2019a). Or, in the words of Russell-Rose et al., based on Tait (2014):

> Professional search focuses on the work of paid professionals who are undertaking a work task that is predominately search-related and performed under a number of constraints such as budget and time. (Russell-Rose et al., 2018)

Professional search is a relatively small area of research in the IR field, but it has been recognized as an important application domain that is challenging because of the specific needs of professional search engine users (Russell-Rose et al., 2018; Salampasis et al., 2013; Verberne et al., 2018b, 2019b). Personalization of search results could potentially increase the effectiveness of professional search, but when developing these methods, we should be aware that professional users – more than users in generic Web search – need to be in control of the search process and must be able to trust the system to provide them with reliable information. Thus, transparency of the retrieval system is essential in this context.

This chapter addresses the user perspective in professional search. In Section 17.2 we introduce professional search as a research area. In Section 17.3 we discuss relevant work on personalization in information retrieval. We then summarize the recent studies addressing explainable search and recommendation in Section 17.4, and in Section 17.5 we give an outline for research directions in the near future, aiming at explainable professional search that makes the user perspective central in the search process.

17.2 Professional Search

Professional searchers, such as lawyers, information specialists, policy officers, architects, and scholars, need to process increasing amounts of documents to find relevant, complete, high-quality, work-related information (Bawden and Robinson, 2009; Sappelli, 2016).

In the common Web search paradigm, as implemented by search engines such as Google and DuckDuckGo, result ranking largely relies on popularity of Web pages: the more hyperlinks from popular pages link to a document, and the more often a document is clicked for a given query, the higher it is ranked in future searches (Joachims et al., 2005). For example, soon after Wikipedia became popular on the Web, Wikipedia pages started to end up on the top of the list on Google result pages for many queries – and this is still the case (DuckDuckGo, 2020). In day-to-day Web search, many users have the same information needs, and therefore popularity is a relevant ranking criterion.

A problem that arises when this popularity-driven paradigm is applied to work-related search is that the most popular documents are often not the most relevant documents for the individual user in their current search task. The differences between professional information search and generic Web search can be summarized in three important aspects:

- The search tasks of professionals are complex, that is, highly specific and typically recall-oriented: the searchers want to be sure that they have found all the relevant information (Kim et al., 2011; Mason, 2006);
- The searching is not limited to sending one query and clicking one result, but is often exploratory by nature (He et al., 2013), and includes browsing, analyzing (Makri et al., 2008), and re-finding previously used information (Sappelli et al., 2017);
- Each user has their own individual needs: not only interests, expertise, and information needs differ per user, but also the perceived relevance of retrieved documents (Sun et al., 2008). The search evolves on the searcher's own knowledge.

Because the information needs in professional search are highly specific and individual, the relevance of the results depends heavily on the *user perspective*. Therefore, the click data available from other users is limited and irrelevant (Huang et al., 2016). Hence, result ranking cannot depend on popularity. An alternative is to use the searcher's own history for improving the search ranking. To achieve that, a user profile must be created and utilized for personalized ranking (Micarelli and Sciarrone, 2004). This brings us to the topic of personalized IR.

17.3 Personalized IR

User profiling and personalization have been addressed extensively in IR research (Ghorab et al., 2013). Approaches to user profiling and personalization typically learn user preferences by collecting queries and clicked documents (Micarelli et al., 2007). A rich user profile can be learned by extracting prominent terms from the clicked documents and storing them in a term profile (Tang et al., 2010; Teevan et al., 2005). Often, the extracted terms are connected to an existing domain knowledge base, for example, a legal thesaurus or a medical ontology (Daoud et al., 2009; Speretta and Gauch, 2005). The term "profile" can then be used to better help the user find relevant information.

One way to do that is to re-order the results based on similarity to the user profile, where the documents that are in the interest field of the user are ranked higher (Micarelli and Sciarrone, 2004). For identifying which documents are relevant to the user profile, and which are not, it is sometimes necessary to perform *query disambiguation* (Tanudjaja and Mui, 2002). Queries are often short, and the user has a specific underlying intent in mind that is unknown to the search engine. A classic example is the query "java," which can refer to either the island or the programming language. The user profile can help in deciding which of the two meanings is more of interest to the user.

The user profile can also be used for personalized query expansion (Zhou et al., 2012) – expanding the user with relevant terms based on the domain of the user, or query suggestion (Leung et al., 2008; Verberne et al., 2015) – showing terms to the user that are likely to be relevant additions to the query. For example, when I enter the query "search behaviour" in Google Scholar (see Figure 17.1), the first three results are about marine predator search behavior, visual search behavior in expert soccer goalkeepers, and job search behavior. Of course, as an IR researcher, I am interested in *information* search behavior, and adding the word "information" to my query improves the relevance of the search results. A query *suggestion* module could detect this and help me improve my query; a query *expansion* module could in the background compare my query with my user profile and add user-specific terms to my query automatically.

Although all research cited here reports an improvement of personalization over the non-personalized baseline, the actual implementation of personalization strategies in search environments is limited: on average, only 11.7% of Google Web Search results show differences due to personalization (Hannak et al., 2013; Hannák et al., 2017). This is because users are wary when it comes to personalization; they feel that their privacy is violated when the search engine uses their personal information. Privacy-preserving personalization is

[HTML] Scaling laws of marine predator **search behaviour**
DW Sims, EJ Southall, NE Humphries, GC Hays... - Nature, 2008 - nature.com
Many free-ranging predators have to make foraging decisions with little, if any, knowledge of present resource distribution and availability 1. The optimal **search** strategy they should use to maximize encounter rates with prey in heterogeneous natural environments remains a ...
☆ 99 Cited by 849 Related articles All 35 versions

Anticipation and visual **search behaviour** in expert soccer goalkeepers
GJP Savelsbergh, J Van der Kamp, AM Williams... - Ergonomics, 2005 - Taylor & Francis
A novel methodological approach is presented to examine the visual **search** behaviours employed by expert goalkeepers during simulated penalty kick situations in soccer. Expert soccer goalkeepers were classified as successful or unsuccessful based on their ...
☆ 99 Cited by 358 Related articles All 14 versions

Networking as a job **search behaviour**: A social network perspective
G Van Hoye, EAJ Van Hooft... - Journal of Occupational ..., 2009 - Wiley Online Library
Although networking is typically recommended as a job **search** strategy in the popular press, research on networking as a job **search behaviour** is scarce. On the basis of social network theory, the present study investigated whether the structure and composition of job seekers' ...
☆ 99 Cited by 256 Related articles All 18 versions

(a)

search behaviour

information search behaviour
job search behaviour
consumer search behaviour
visual search behaviour
marine predator search behaviour
search behaviour **scaling laws**
web search behaviour
search behaviour **unemployment**

Anticipation and visual **search behaviour** in expert soccer goalkeepers
GJP Savelsbergh, J Van der Kamp, AM Williams... - Ergonomics, 2005 - Taylor & Francis
A novel methodological approach is presented to examine the visual **search** behaviours employed by expert goalkeepers during simulated penalty kick situations in soccer. Expert soccer goalkeepers were classified as successful or unsuccessful based on their ...
☆ 99 Cited by 358 Related articles All 14 versions

Networking as a job **search behaviour**: A social network perspective
G Van Hoye, EAJ Van Hooft... - Journal of Occupational ..., 2009 - Wiley Online Library
Although networking is typically recommended as a job **search** strategy in the popular press, research on networking as a job **search behaviour** is scarce. On the basis of social network theory, the present study investigated whether the structure and composition of job seekers' ...
☆ 99 Cited by 256 Related articles All 18 versions

(b)

information search behaviour

About 4.420.000 results (0,05 sec)

The effect of prior destination experience on online **information search behaviour**
XY Lehto, DY Kim, AM Morrison - Tourism and Hospitality ..., 2006 - journals.sagepub.com
This research proposed and empirically tested a behavioural model of online tourist **information search** by integrating the consumer involvement concept, modes of cognitive processing and knowledge representation. The model identified two types of online ...
☆ 99 Cited by 216 Related articles All 6 versions

A study of the **information search behaviour** of the millennial generation.
A Taylor - **Information** research: an international electronic journal, 2012 - ERIC
Introduction. Members of the millennial generation (born after 1982) have come of age in a society infused with technology and **information**. It is unclear how they determine the validity of **information** gathered, or whether or not validity is even a concern. Previous **information** ...
☆ 99 Cited by 112 Related articles All 5 versions

A review of factors affecting online consumer **search behaviour** from an **information** value perspective
R Grant, RJ Clarke, E Kyriazis - Journal of Marketing Management, 2007 - Taylor & Francis
The use by consumers of the internet as a means of purchasing products and services seems to have fallen short of expectations that arose with consumer access to the internet. Evidence from online sales continues to show low conversion ratios, from enquiry to sale, in ...
☆ 99 Cited by 153 Related articles All 5 versions

(c)

Figure 17.1 Example of personal relevance in academic search (Google Scholar). (a) The query "search behaviour" gives results that are irrelevant for me; (b) Google's query suggestion functionality suggests specifications of my query. The first suggestion is relevant to me; (c) If I search for "information search behaviour" I receive relevant results. Google and the Google logo are trademarks of Google LLC

therefore an important societal topic (Karwatzki et al., 2017; Mittelstadt, 2016). A crucial step in the development of privacy-secure systems is to make the system transparent and explainable to the user (Holzinger et al., 2017). This is further elaborated in the next section.

Transparency and explainability are even more important in work-related contexts than in the Web search context: professional users do not want to feel that they are losing control over the search process because the ranking of the search results is not stable or not predictable (Russell-Rose et al., 2018).

17.4 Explainable Recommendation and Search

In artificial intelligence, explainability is an important means to address issues with transparency and trust in black-box machine learning models (Adadi and Berrada, 2018). Search and recommender systems are ubiquitous in our daily lives, and it is considered important that users understand the results and recommendations they receive (Zhang et al., 2019). Recommender systems such as Amazon, Spotify, and Netflix recommend items to users that they are likely to appreciate (buy, listen, watch), without waiting for a user's query. Search systems provide information that is estimated to be relevant to a query entered by the user. In both types of systems, it can be valuable for the user to see why information was presented to them. In explainable *personalized* search, the user gets an answer to the question: "Why was this information presented to me," and even "Why should I trust the given information?"

Explainability in Recommender Systems

Early approaches to recommendation were inherently explainable because they were relatively straightforward, directly related to the content of the suggested items and their user ratings. Examples of explainable recommendations in these systems are "You have highly rated items that are similar to this item" and "Users who have similar ratings with you highly rated this item." With the rise of machine learning methods in recommender systems for a large audience (e.g., Netflix, Spotify), the implementation of explainability and its trade-off with accuracy became more challenging (Koren et al., 2009; Zhang et al., 2014). The trade-off is based on the discrepancy that machine learning systems, and in particular deep neural network approaches, are outperforming rule-based systems in the quality of recommended results, but they are much more difficult to explain to the user than simple rule-based systems.

In recent years, the use of knowledge graphs to facilitate explainability recommendation methods has been proposed (Ai et al., 2018; Wang et al., 2018). Knowledge graphs are flexible and can integrate heterogeneous information types. Users and items are both modeled as nodes (entities) in the graph and the strengths of the relations between entities are used for recommending new entities to a user. Explanations can be generated in natural language to explain the relevance of a specific item to the user (Balog et al., 2019).

Explainability in Search Systems

Personalized search has in common with recommender systems that the user profile determines the relevance of a document. But as opposed to recommender systems, retrieval systems get an input query and need to retrieve documents that are relevant to that particular query.

In explainable search, the aim is that the user knows the capabilities and limitations of the search system, that they trust the system, and know how to intervene with the system if the results are not satisfactory (Zhang et al., 2019). A common form of explanation of the relevance of search results is the use of search snippets on the result page in which query terms have been marked in boldface. This markup (which is used in all Web search engines) indicates the topical relevance of a document for the user query. Search snippets are a basic example of explanations unified with the ranking model: Ranking models have term overlap metrics as central components, and term overlap is directly visualized in the snippets on the result page.

For relevance factors that, as opposed to term overlap, do not directly follow from the ranking model, a separate explanation engine is needed that generates explanations post hoc. This need has become more urgent as the state of the art in IR is now held by deep neural network models that do not use human-defined features for ranking the documents but abstracted document representations (Nogueira and Cho, 2019; Yang et al., 2019). Since 2019, efforts have been made to make features and their importance weights from neural retrieval models explicit and visualize these features as an explanation of why a document is relevant to the user query (Chios and Verberne, 2020; Fernando et al., 2019; Singh and Anand, 2019).

In the context of *professional* search, explainability is a novel research direction with no experimental results published yet at the time of writing. In the next section, we list suggestions for research in this direction.

17.5 Toward Explainable Professional Search

After having discussed professional search (Section 17.2), personalized search (Section 17.3), and explainable search (Section 17.4) in the previous three sections, we now bring these topics together in the next step for advancing professional search: the development of explainable search methods in the professional context that allow for personalization without becoming a black box to the user.

According to Russell-Rose and MacFarlane (2020 p. 2), explainability in professional search has two criteria: (1) the ease with which the user's information need can be translated into a query (explainability of the query process), and (2) the degree to which the user's query returns the results expected and intended by the user (explainability of the search results). Explainability of the query process strongly relates to professional query interfaces, which often allow the user to build complex Boolean queries. Russell-Rose and MacFarlane recommend improving the explainability of query interfaces by providing real-time feedback on query effectiveness, allowing users to evaluate the contribution of individual query elements (Russell-Rose and MacFarlane, 2020; Russell-Rose and Shokraneh, 2020, p. 4).

In this chapter, we focus on the second criterion: the explainability of the search results.

Since professional search tasks are highly specific, result ranking cannot rely on the data of other users. Given this individual nature of professional search, personalization of professional search seems a logical step. However, when using user information in relevance ranking, it is important for users to have insight in to the data that is stored by the search engine (Xu et al., 2007) and to understand the influence of their personal data on the search results. Thus, professional search relies on explainable models in order to have the user trust the system and be in control of the search process.

The current state of research, discussed in the previous sections, gives way to two research directions for the near future:

1. Post hoc explanations added to the ranked lists of documents;
2. Graph-based personalized search, explicitly adding the individual user perspective to the searching and browsing process.

17.5.1 Explanations for Estimated Document Relevance

Just as snippets give an indication of topical relevance by highlighting query terms in text excerpts, other relevance factors could also be explicitly highlighted on the result page. These relevance factors could differ between domains. For example, users of a legal search engine (lawyers, legal scholars, legal professionals) consider document characteristics such as source authority, legal hierarchy, and whether the document is annotated to be important factors of relevance (Wiggers et al., 2018). Adding such metadata information of the retrieved documents to the result page is relatively straightforward; the next step would be to show indications of the weight that the ranking model assigned to each relevance factor. This helps inform the user about which factors were taken into account for the ranking and how they were weighted. A paper by Chios and Verberne (2020) proposed a search engine result page on which the relative importance of query terms for the retrieved documents and the position of the most relevant passage in each document are shown. This was positively valued by the participants of a small-scale user study: they give significantly higher scores for the explainability and assessability (how well can the relevance of the retrieved documents be assessed) of the result page. This paper could be followed up by work addressing relevance factors in professional contexts.

17.5.2 Explainable Search Using a Personal Graph

A promising direction for explainable search is the use of graph models. Graphs are a natural and transparent means of representing knowledge (Chein and Mugnier, 2008). Knowledge graphs have been shown to be especially helpful in exploratory search tasks (Sarrafzadeh et al., 2014, 2016), which are common in professional work environments (He et al., 2013). Graphs can also be used for generating search explanations by explicitly describing the path between users and items in the graph. If we take academic search (Chiang et al., 2013; Salehi et al., 2015; Verberne et al., 2015, 2018b) as an example, we could generate explanations such as "this article is retrieved because you have previously read papers that cite it, and because you commonly read papers from this journal."

Most previous works in graph-based search use an external knowledge graph covering all domain knowledge. Verberne (2018) and Balog and Kenter (2019) have both proposed storing personal knowledge graphs to enable personalized search. A personal knowledge graph is "a resource of structured information about entities personally related to its user, their attributes and the relations between them" (Balog and Kenter, 2019). Thus, the personal knowledge graph is a possible visualization of the user's perspective in the search process. In the proposal by Verberne (2018), the personal knowledge graph is a professional graph representing the searching and browsing history of the user in the professional search engine. A graph representing the knowledge and interests of one user is much smaller than a graph representing the complete index of a search engine (Blanco and Lioma, 2012) and can be stored locally (client-side), if privacy regulations require it.

There are two main challenges associated with the idea of a personal graph for information search: automatically populating the personal graph from sparse user data and effectively utilizing the graph for effective information finding.

Future research with professional knowledge graphs should address the development of methods for these two aspects.

17.6 Conclusions

In this chapter, we have discussed the idea of the Perspective Web in the context of IR, and in particular information search for professional purposes. In their search for information, users act based on their own, personal perspective. User queries in Web search engines are often underspecified because much of the user context is implicit. The underspecificy of user queries leads to ambiguity: Does the query term "search behavior" refer to *predator* search behavior, *job* search behavior, or *information* search behavior? The user knows, but the search engine does not. Modern Web search engines solve this by showing a diversity of perspectives to the user, hoping that a relevant result is among these. In the ranking of results in Web search engines, popularity is an important criterion: the more users have clicked on a Web page, the more often it shows up in the result list of other users.

Professional search tasks are user- and context-specific. This means that the user perspective plays an even larger role in the relevance of the returned results than in Web search; marine predator search behavior would be relevant from the perspective of a marine biologist, but not from the perspective of me as an IR researcher. Thus, ranking algorithms cannot use the popularity of search results as effectively as in Web search. This establishes the potential for personalization in professional search. At the same time, professional users want to be in control over the search process and need to be able to trust the search engine to provide them with correct and relevant information. This motivates the necessity to make the results retrieved by the professional search engine explainable to the user. We have discussed the state of the art in explainable recommendation and search and then proposed two possible research directions for explainable professional search: adding explanations to traditional result pages and developing a search paradigm that is centered around the user's professional knowledge graph. This latter research direction could potentially lead to a true realization of the user perspective in information search.

There is one caveat to a search engine that centers around the user's perspective, and that is the filter bubble effect (Nguyen et al., 2014): if the user profile is based on the user's past behavior and the user profile is used to change the search results, the risk is that the user will dive deeper in directions that confirm their own beliefs (perspective), ignoring the results that contradict them. This is one of the reasons why user control is important: the user needs to see at any time what the influence of their user profile is on the results they see.

The explainable interface needs to include visual information on the user perspective itself. In my vision, the graph visualization proposed in Section 17.5.2 would become a kaleidoscope where a different perspective changes the view of the data.

References

Adadi, A., and Berrada, M. 2018. Peeking inside the black-box: A survey on Explainable Artificial Intelligence (XAI). *IEEE Access*, **6**, 52138–52160.

Ai, Q., Azizi, V., Chen, X., and Zhang, Y. 2018. Learning heterogeneous knowledge base embeddings for explainable recommendation. *Algorithms*, **11**(9), 137.

Balog, K., and Kenter, T. 2019. Personal knowledge graphs: A research agenda. Pages 217–220 of: *Proceedings of the 2019 ACM SIGIR International Conference on Theory of Information Retrieval*. ICTIR '19. New York: ACM.

Balog, K., Radlinski, F., and Arakelyan, S. 2019. Transparent, scrutable and explainable user models for personalized recommendation. Pages 265–274 of: *Proceedings of the 42nd International ACM SIGIR Conference on Research and Development in Information Retrieval*. SIGIR '19. New York: ACM.

Bawden, D., and Robinson, L. 2009. The dark side of information: Overload, anxiety and other paradoxes and pathologies. *Journal of Information Science*, **35**(2), 180–191.

Blanco, R., and Lioma, C. 2012. Graph-based term weighting for information retrieval. *Information Retrieval*, **15**(1), 54–92.

Chein, M., and Mugnier, M.-L. 2008. *Graph-based knowledge representation: computational foundations of conceptual graphs*. London: Springer Science & Business Media.

Chiang, M.-F., Liou, J.-J., Wang, J.-L., Peng, W.-C., and Shan, M.-K. 2013. Exploring heterogeneous information networks and random walk with restart for academic search. *Knowledge and Information Systems*, **36**(1), 59–82.

Chios, I., and Verberne, S. 2020. Helping results assessment by adding explainable elements to the deep relevance matching model. Pages 1–9 of *Proceedings of the 3rd International Workshop on ExplainAble Recommendation and Search*. Xi'an: EARS '20.

Croft, W. B. 2019. The importance of interaction for information retrieval. Pages 1–2 of: *Proceedings of the 42nd International ACM SIGIR Conference on Research and Development in Information Retrieval*. SIGIR '19. New York: ACM.

Daoud, M., Lechani, L.-T., and Boughanem, M. 2009. Towards a graph-based user profile modeling for a session-based personalized search. *Knowledge and Information Systems*, **21**(3), 365–398.

DuckDuckGo 2020. *Information retrieval at DuckDuckGo*. https://duckduckgo.com/?t=ffab&q=information+retrieval&ia=web.

Fernando, Z. T., Singh, J., and Anand, A. 2019. A study on the interpretability of neural retrieval models using DeepSHAP. Pages 1005–1008 of: *Proceedings of the 42nd International ACM SIGIR Conference on Research and Development in Information Retrieval*. Paris: ACM.

Ghorab, M. R., Zhou, D., O'Connor, A., and Wade, V. 2013. Personalised information retrieval: Survey and classification. *User Modeling and User-Adapted Interaction*, **23**(4), 381–443.

Hannák, A., Sapieżyński, P., Khaki, A. M., Lazer, D., Mislove, A., and Wilson, C. 2017. Measuring personalization of Web search. *arXiv preprint arXiv:1706.05011*.

Hannak, A., Sapiezynski, P., Molavi Kakhki, A., Krishnamurthy, B., Lazer, D., Mislove, A., and Wilson, C. 2013. Measuring personalization of Web search. Pages 527–538 of: *Proceedings of the 22nd International Conference on World Wide Web*. WWW '13. New York: ACM.

Harman, D. 2019. Information retrieval: The early years. *Foundations and Trends® in Information Retrieval*, **13**(5), 425–577.

He, J., Bron, M., and de Vries, A. P. 2013. Characterizing stages of a multi-session complex search task through direct and indirect query modifications. Pages 897–900 of: *Proceedings of the 36th International ACM SIGIR Conference on Research and Development in Information Retrieval*. Dublin: ACM.

Holzinger, A., Biemann, C., Pattichis, C. S., and Kell, D. B. 2017. What do we need to build explainable AI systems for the medical domain? *arXiv preprint arXiv:1712.09923*.

Huang, Z., Cautis, B., Cheng, R., and Zheng, Y. 2016. KB-enabled query recommendation for long-tail queries. Pages 2107–2112 of: *Proceedings of the 25th ACM International on Conference on Information and Knowledge Management*. Indianapolis: ACM.

Ingwersen, P., and Järvelin, K. 2005. *The turn: Integration of information seeking and retrieval in context (The information retrieval series), Vol. 18*. Dordrecht: Springer.

Joachims, T., Granka, L., Pan, B., Hembrooke, H., and Gay, G. 2005. Accurately interpreting clickthrough data as implicit feedback. Pages 154–161 of: *Proceedings of the 28th Annual International ACM SIGIR Conference on Research and Development in Information Retrieval*. New York: ACM.

Karwatzki, S., Dytynko, O., Trenz, M., and Veit, D. 2017. Beyond the personalization–privacy paradox: Privacy valuation, transparency features, and service personalization. *Journal of Management Information Systems*, **34**(2), 369–400.

Kim, Y., Seo, J., and Croft, W. B. 2011. Automatic Boolean query suggestion for professional search. Pages 825–834 of: *Proceedings of the 34th International ACM SIGIR Conference on Research and Development in Information Retrieval*. SIGIR '11. New York: ACM.

Koren, Y., Bell, R., and Volinsky, C. 2009. Matrix factorization techniques for recommender systems. *Computer*, **42**(8), 30–37.

Leung, K. W.-T., Ng, W., and Lee, D. L. 2008. Personalized concept-based clustering of search engine queries. *IEEE Transactions on Knowledge and Data Engineering*, **20**(11), 1505–1518.

Makri, S., Blandford, A., and Cox, A. L. 2008. Investigating the information-seeking behaviour of academic lawyers: From Ellis's model to design. *Information Processing & Management*, **44**(2), 613–634.

Mason, D. 2006. Legal information retrieval study–Lexis professional and Westlaw UK. *Legal Information Management*, **6**(4), 246–250.

Maxwell, D. M. 2019. Modelling search and stopping in interactive information retrieval. Ph.D. thesis, University of Glasgow.

Micarelli, A., Gasparetti, F., Sciarrone, F., and Gauch, S. 2007. Personalized search on the World Wide Web. *The Adaptive Web*, **4321**, 195–230.

Micarelli, A., and Sciarrone, F. 2004. Anatomy and empirical evaluation of an adaptive Web-based information filtering system. *User Modeling and User-Adapted Interaction*, **14**(2), 159–200.

Mittelstadt, B. 2016. Auditing for transparency in content personalization systems. *International Journal of Communication*, **10**(June), 4991–5002.

Nguyen, T. T., Hui, P.-M., Harper, F. M., Terveen, L., and Konstan, J. A. 2014. Exploring the filter bubble: The effect of using recommender systems on content diversity. Pages 677–686 of: *Proceedings of the 23rd International Conference on World Wide Web*. Seoul: ACM.

Nogueira, R., and Cho, K. 2019. Passage re-ranking with BERT. *arXiv preprint arXiv:1901.04085*.

Russell-Rose, T., Chamberlain, J., and Azzopardi, L. 2018. Information retrieval in the workplace: A comparison of professional search practices. *Information Processing & Management*, **54**(6), 1042–1057.

Russell-Rose, T., and MacFarlane, A. 2020. Towards explainability in professional search. *Proceedings of the 3rd International Workshop on ExplainAble Recommendation and Search*. EARS '20.

Russell-Rose, T., and Shokraneh, F. 2020. Designing the structured search experience: Rethinking the query-builder paradigm. *Weave: Journal of Library User Experience*, **3**(1).

Salampasis, M., Fuhr, N., Hanbury, A., Lupu, M., Larsen, B., and Strindberg, H. 2013. Integrating IR technologies for professional search. Pages 882–885 of: Serdyukov, P., Braslavski, P., Kuznetsov, S. O., Kamps, J., Rüger, S., Agichtein, E., Segalovich, I., and Yilmaz, E. (eds), *Advances in information retrieval*. Berlin: Springer.

Salehi, S., Du, J. T., and Ashman, H. 2015. Examining personalization in academic Web search. Pages 103–111 of: *Proceedings of the 26th ACM Conference on Hypertext & Social Media*. Guzelyurt: ACM.

Sappelli, M. 2016. Knowledge work in context. User centered knowledge worker support. Ph.D. thesis, Radboud University Nijmegen.

Sappelli, M., Verberne, S., and Kraaij, W. 2017. Evaluation of context-aware recommendation systems for information re-finding. *Journal of the Association for Information Science and Technology*, **68**(4), 895–910.

Sarrafzadeh, B., Vechtomova, O., and Jokic, V. 2014. Exploring knowledge graphs for exploratory search. Pages 135–144 of: *Proceedings of the 5th Information Interaction in Context Symposium on - IIiX '14*. Regensburg: ACM.

Sarrafzadeh, B., Vtyurina, A., Lank, E., and Vechtomova, O. 2016. Knowledge graphs versus hierarchies. Pages 91–100 of: *Proceedings of the 2016 ACM on Conference on Human Information Interaction and Retrieval – CHIIR '16*. Carrboro: ACM.

Singh, J., and Anand, A. 2019. EXS: Explainable search using local model agnostic interpretability. Pages 770–773 of: *Proceedings of the Twelfth ACM International Conference on Web Search and Data Mining*. Melbourne: ACM.

Speretta, M., and Gauch, S. 2005. Personalized search based on user search histories. Pages 622–628 of: *Web Intelligence, 2005. Proceedings: The 2005 IEEE/WIC/ACM International Conference on Web Intelligence.*

Sun, Y., Li, H., Councill, I. G., Huang, J., Lee, W.-C., and Giles, C. L. 2008. Personalized ranking for digital libraries based on log analysis. Pages 133–140 of: *Proceedings of the 10th ACM workshop on Web Information and Data Management.* Napa Valley: ACM.

Tait, J. I. 2014. An introduction to professional search. Goos, G., Hartmanis, J., and van Leeuwen, J. (eds.) Pages 1–5 of: *Professional search in the modern world.* Springer.

Tang, J., Yao, L., Zhang, D., and Zhang, J. 2010. A combination approach to Web user profiling. *ACM Transactions on Knowledge Discovery from Data (TKDD)*, **5**(1), 2.

Tanudjaja, F., and Mui, L. 2002. Persona: A contextualized and personalized Web search. Pages 1232–1240 of: *Proceedings of the 35th Annual Hawaii International Conference on System Sciences.* Big Island: IEEE.

Teevan, J., Dumais, S. T., and Horvitz, E. 2005. Personalizing search via automated analysis of interests and activities. Pages 449–456 of: *Proceedings of the 28th Annual International ACM SIGIR Conference on Research and Development in Information Retrieval.* Salvador: ACM.

Verberne, S. 2018. Explainable IR for personalizing professional search. Pages 35–42 of: *Joint Proceedings of the First International Workshop on Professional Search (ProfS2018); the Second Workshop on Knowledge Graphs and Semantics for Text Retrieval, Analysis, and Understanding (KG4IR); and the International Workshop on Data Search.* Ann Arbor: SEARCH'18.

Verberne, S., He, J., Kruschwitz, U., Larsen, B., Russell-Rose, T., and de Vries, A. P. 2018b. First international workshop on professional search (ProfS2018). Pages 1431–1434 of: *The 41st International ACM SIGIR Conference on Research & Development in Information Retrieval.* SIGIR '18. New York: ACM.

Verberne, S., He, J., Kruschwitz, U., Wiggers, G., Larsen, B., Russell-Rose, T., and de Vries, A. P. 2019a. First international workshop on professional search. Pages 153–162 of: *ACM SIGIR Forum*, Vol. 52. New York: ACM.

Verberne, S., He, J., Wiggers, G., Russell-Rose, T., Kruschwitz, U., and de Vries, A. P. 2019b. Information search in a professional context-exploring a collection of professional search tasks. *arXiv preprint arXiv:1905.04577.*

Verberne, S., Kraaij, W., and de Vries, A. P. 2018a. Author-topic profiles for academic search. *arXiv preprint arXiv:1804.11131.*

Verberne, S., Sappelli, M., Kalervo, J., Järvelin, K., and Kraaij, W. 2015. User simulations for interactive search: Evaluating personalized query suggestion. Pages 678–690 of: *Advances in information retrieval, Vol. 9022.* Vienna: Springer International.

Wang, H., Zhang, F., Wang, J., Zhao, M., Li, W., Xie, X., and Guo, M. 2018. Ripplenet: Propagating user preferences on the knowledge graph for recommender systems. Pages 417–426 of: *Proceedings of the 27th ACM International Conference on Information and Knowledge Management.* Torino: ACM.

Wiggers, G., Verberne, S., and Zwenne, G.-J. 2018. Exploration of intrinsic relevance judgments by legal professionals in information retrieval systems. Pages 5–8 of: *Proceedings of the 17th Dutch-Belgian Information Retrieval Workshop.* Leiden: arXiv.

Xu, Y., Wang, K., Zhang, B., and Chen, Z. 2007. Privacy-enhancing personalized Web search. Pages 591–600 of: *Proceedings of the 16th International Conference on World Wide Web*. Banff: ACM.

Yang, W., Zhang, H., and Lin, J. 2019. Simple applications of bert for ad hoc document retrieval. *arXiv preprint arXiv:1903.10972*.

Zhang, Y., Lai, G., Zhang, M., Zhang, Y., Liu, Y., and Ma, S. 2014. Explicit factor models for explainable recommendation based on phrase-level sentiment analysis. Pages 83–92 of: *Proceedings of the 37th International ACM SIGIR Conference on Research & Development in Information Retrieval*. Gold Coast: ACM.

Zhang, Y., Mao, J., and Ai, Q. 2019. SIGIR 2019 tutorial on explainable recommendation and search. Pages 1417–1418 of: *Proceedings of the 42nd International ACM SIGIR Conference on Research and Development in Information Retrieval*. Paris: ACM.

Zhou, D., Lawless, S., and Wade, V. 2012. Improving search via personalized query expansion using social media. *Information Retrieval*, **15**(3–4), 218–242.

18
Harvesting Perspectives in Social Media

Tommaso Caselli and Malvina Nissim

18.1 Introduction

Polarization[1] of the public debate is an aspect that appears to characterize more and more interactions among individuals both in real life and on social media platforms. Problems may arise when such polarized interactions become disruptive of the debate or debase participants. Manually monitoring all online interactions is not possible, even for big corporations like Facebook or Twitter. The volume of data requires that ways are found to classify online content automatically (Kennedy et al., 2017; Nobata et al., 2016) so as to assist humans in monitoring and reviewing such content.

A requirement to move in this direction is the availability of datasets to train dedicated systems. Recently, there has been indeed an increased interest in developing datasets that address different aspects of this phenomenon, including rumors and fake news (Derczynski et al., 2017; Gorrell et al., 2019; Hanselowski et al., 2018), controversy (Borra et al., 2015; Coletto et al., 2017; Dori-Hacohen and Allan, 2015; Garimella et al., 2016; Mejova et al., 2014), toxic language (van Aken et al., 2018; Wulczyn et al., 2016), offensive and/or hateful speech (Basile et al., 2019; Founta et al., 2018; Schmidt and Wiegand, 2017; Zampieri et al., 2019), among others. This explosion of perspective-oriented data resulted in the development of numerous supervised systems. Looking at this process as an assembly line (annotated data, models, results), we identify in its very first block a huge critical issue. As many contributions have pointed out (Geva et al., 2019; Gururangan et al., 2018; Paun et al., 2018; Tsuchiya, 2018), manually annotated data has limitations with respect to different aspects. First, size: manually annotated data will always yield small datasets not necessarily sufficient to optimally train models. Second, data selection, that

[1] This contribution is based on three previously published works by the same authors with other collaborators. We point to each relevant publication when necessary.

is, what to keep and what to exclude, can introduce bias with respect to a specific topic or incidence of a phenomenon. For instance, Wiegand et al. (2019) have shown that one popular dataset for hate speech, namely Wassem & Hovy (WH) (Waseem and Hovy, 2016), has a big topic bias because some of the selected messages were taken from a popular kitchen TV show, resulting in models that would consider terms like "kitchen," "table," and similar as good indicators of hate speech. Third, annotators, both experts and/or crowd-based, can introduce personal biases in the annotations which do not necessarily reflect the intentions of the author of the message (i.e., the writer) and the goal of the annotation task. All of these problems are exacerbated by the high subjective nature of the tasks.

In this contribution, we illustrate three methods to investigate different perspective-related phenomena in Social Media, namely **controversies**, **hate speech**, and **abusive language**, aiming to avoid or address the shortcomings we have illustrated. We promote approaches inspired by the paradigm of *distant supervision*, showing how by using specific *proxies* we can target different phenomena. We validate our proposals through various experiments. Our experiments are either fully ecological, implying no manual annotation of the data, or partially ecological, meaning that we use a combination of distantly labeled and manually annotated data.

This chapter offers the following contributions:

- a series of distantly supervised methods for harvesting perspectives and polarized content in Social Media. The advantage of such methods is that, while noisy, they stay closer to the data than any over-imposed annotation strategy done by others;
- a series of case studies to test such methods, covering different downstream tasks, namely controversy detection, hate-speech detection, and abusive language in general. The experiments are run on multiple languages to show that their applicability is not language dependent, provided that access to data is available;
- reflections over both methodology and results with insights and open issues for future work.

18.2 Methods

Harvesting perspectives through manual annotation runs the risk of bias at different levels: bias can affect the annotation guidelines, the annotators, even when sufficiently trained, as well as the documents that compose the corpus. To avoid such risks, we promote a more natural approach based on *distant supervision*. Under its general paradigm, distant supervision requires that the training material is identified through so-called silver labels that can be identified either by means of heuristics or by exploiting clues in the data, called *proxies*, rather than manually annotated data. Distant supervision has been successfully applied in different areas such as relation extraction (Mintz et al., 2009), semantic role labeling (Exner et al., 2015), sentiment analysis (Go et al., 2009), and emotion detection (Pool and Nissim, 2016).

To harvest perspectives in social media in a distantly supervised approach, we propose three methods which in practice translate into three kinds of proxies:

- the use of the *distribution* of Facebook reactions to posts;
- the use of Facebook pages, which we call *sources*;
- the use of *hashtags* from Twitter.

We apply and show the effectiveness of these methods in three different tasks, all falling under the wider umbrella of "perspective" in social media, each capturing a different facet of this phenomenon. We explore the portability of these methods by applying them across languages and domains. Specifically, we targeted the following perspective-related problems:

- detecting controversial news using Facebook, in Italian and Dutch;
- detecting hate speech in Facebook and Twitter messages in Italian and German;
- detecting abusive language in online messages in English.

Entropy for Controversy A controversy is best defined as any lengthy interaction on a topic (e.g., gun control, abortion, among others) where participants' opinions remain unchanged, and become more and more polarized toward extreme values (Timmermans et al., 2017).

Facebook reactions extend the users' possibilities to express an "opinion" on a post by choosing among five emotions (`ANGRY`, `HAHA`, `WOW`, `SAD`, and `LOVE`) next to the well-know `LIKE`. On the basis of the definition of controversy previously introduced, our working hypothesis is that if users' reactions fall in two or more emotion classes (not necessarily opposed in terms of "polarity") with high frequencies, the controversy of a news item is higher. To quantify this intuition, we use *entropy*, a measure from Information Theory that expresses the degree of uncertainty of a given variable. In our case, this concerns the distribution of the Facebook reactions: the more dishomogeneous (i.e., the more varied the reactions to a given post), the higher the entropy. To better clarify this aspect, consider the data in Table 18.1. Each sample is the title of a news article in Italian and Dutch from Facebook, for which we report the reactions' breakdown,

Table 18.1 *Entropy variation in relation to the reactions for Italian (1.) and (2.) and Dutch (3.) and (4.)*

Text	LIKE	ANGRY	HAHA	WOW	SAD	LOVE	Entropy
Come in un reality, Trump nomina Neil Gorsuch in diretta web.	0	25	0	0	0	16	0.964
Era la notizia che la famiglia Cucchi attendeva da tempo	5	604	0	0	31	25	0.563
Vliegramp in Iran: 66 doden	107	0	83	1	200	1	0.657
Nederlander is het sparen verleerd	43	22	9	28	10	2	0.941

and the overall entropy based on reaction counts. The more distributed the reactions, the more likely they may signal the probability of a text to be controversial. This is reflected by the entropy scores, whose difference in values suggests varying levels of controversy.[2]

Sources for Hate Speech According to the European Commission against Racism and Intolerance (ECRI), hate speech is any form of "expressions which spread, incite, promote or justify hatred, violence and discrimination against a person or group of persons for a variety of reasons."[3]

For this set of experiments, rather than using any hints from the content itself, we exploit *where* such content has been published, that is, the sources of the messages. This is a special take on distant supervision. The approach relies on previous studies (Bozdag and van den Hoven, 2015; Pariser, 2011; Seargeant and Tagg, 2018) that have shown how online communities aggregate on specific online fora and tend to reinforce themselves by decreasing diversity, distorting information, and polarizing socio-political opinion.

Facebook offers good venues to further explore these claims. In particular, Facebook users can join groups or follow pages of celebrities and organizations where they communicate and interact with other users. These pages become different sources of data, and their content can be used as proxies for generating specialized data to be applied in downstream tasks that require access to some sort of polarized content.

[2] The translations of the examples in Table 18.1 are the following: *Like in a reality show, Trump nominates Neil Gorsuch live on the Web*; *It was the news that the Cucchi family was long waiting*; *Iran air disaster: 66 dead*; *The Dutch have forgotten about saving.*

[3] `www.coe.int/en/web/european-commission-against-racism-and-intolerance/hate-speech-and-violence`.

Table 18.2 *Embedding comparison: Nearest neighbors of hate targets*

Generic Embeddings	Hateful Embeddings
"immigrati" [migrants]	
immigranti [migrants] (0.737)	extracomunitari [non-EU immigrant] (0.841)
emigranti [emigrants] (0.731)	immigranti [migrants](0.828)
emigrati [emigrated] (0.725)	clandestini [illegals] (0.823)
"trans" [trans]	
europ [europ] (0.399)	lesbo [lesbians] (0.720)
express [express] (0.352)	puttane [whores] (0.709)
airlines [airlines] (0.327)	gay [gay] (0.703)

Focusing on hate speech detection, we identified a set of Facebook pages known to promote or be the target of hate speech. We scraped from these pages comments of users, but rather than using this data directly as training instances, we used it to generate what we call *hate embeddings* to be further integrated into hate speech detection systems. We generated the hate embeddings using the `word2vec` skip-gram model (Mikolov et al., 2013). Table 18.2 illustrates the differences in the most similar lexical items, that is, the nearest neighbors, for words that can be used to identify targets of hate speech between generic word embeddings for Italian and the newly created hate embeddings. Differences in the neighbors sets affect both the connotative aspects of word meaning (i.e., neutral vs. negative connotations for both targets) as well as their semantic area. For instance, the word "trans" [trans] in the generic embedding space appears to be more similar to words related to transportation, as it may be considered an abbreviation. On the other hand, the hate embeddings highlight its similarity/relatedness to the LGBTQ+ communities, as well as negative stereotypes (i.e., transgender people as "puttane" [whores]). Similarly, the word "immigrati" [migrants] in the hate embeddings is more similar/related to words that carry a negative connotation ("extracomunitari" [non-EU immigrant], and "clandestini" [illegals]).

Hashtags for Abusive Language Abusive language is a broad category defined as "[a]ny strongly impolite, rude or hurtful language using profanity, that can show a debasement of someone or something, or show intense emotion" (Founta et al., 2018, page 495).

Direct access to sources is not always possible, as it depends on the features of the various social media platforms. In the case of Twitter, the platform is structured as a network of individuals, and, differently from what happens

Table 18.3 *Embedding comparison: Nearest neighbors of abusive targets*

Generic Embeddings	Polarized Embeddings
immigrant	
immigrants (0.726)	ancstrs (0.804)
undocumented (0.706)	sanctuaries (0.634)
migrant (0.578)	illegal (0.605)
muslim	
muslims (0.7)	deception ☞ every (0.870)
islam (0.748)	pounches (0.859)
islamic (0.651)	terrorize...(0.863)

in Facebook, for instance, groups or communities are not features available to the users. However, previous work (Bryden et al., 2013) has shown that the networks that emerge from user interactions can actually be conceived as a hierarchy of communities. Such communities can be recovered thanks to the words used within the communicative interactions of the networks of individuals. On the basis of these findings, we proxy communities exploiting sets of perspective-bearing hashtags in Twitter. Similarly to the source-based approach, in Table 18.3 we show differences in nearest neighbors between generic Twitter embeddings (Twitter GloVe, Pennington et al. (2014) and the polarized ones for potential targets of abusive language in English. The results mirror those obtained using the source-based approach returning words that are actually more prone to be used in potentially abusive contexts and encoding negative connotations.

18.3 Case Studies

In this section, we illustrate the application of our approaches to collect and generate perspective-loaded data in three different case studies, namely, Controversy Detection (18.3.1), Hate Speech Detection (18.3.2), and Abusive Language Detection (18.3.3). Each task represents a proof-of-concept of the validity of each proposed method, showing strengths and limitations.

18.3.1 Controversy Detection

The prediction of controversy is done by means of a system using a sparse feature representation and a linear regressor, with the *scikit-learn* Linear Regression implementation (Buitinck et al., 2013).

Data for Italian and Dutch were collected with the Facebook Graph API[4] scraping Facebook posts and reactions from four Italian newspapers and one news agency (*Corriere della Sera, La Repubblica, Il Manifesto, il Giornale, ANSA*), and six Dutch newspapers (*Het Parool, RTL Nieuws, NOS, De Volkskrant, De Telegraaf,* and *Het NRC*). Italian data covers a period of and a half two months (mid April–early July 2017), while the Dutch data a period of six months (mid-August 2017, mid-February 2018). We excluded posts with less than 30 reactions in total. We also collected an excerpt of each article (i.e., `text`) and any additional text commenting the article, if available (i.e., `descriptor`). We collected a total of 3,585 posts and almost 2 million reactions for Italian, while 1,859 posts and almost 650,000 reactions for Dutch. In the remainder of this section, we use Italian to describe the experiment. The same procedure was applied to Dutch.

Parameter tuning was conducted on the *ANSA* data using grid search with 10-fold cross-validation. We assume that *ANSA* being a news agency, its texts should be more objective and the potential controversy should depend on the reported event rather than on its framing. We used a tf-idf vectorizer to represent the text as both word and character n-grams. In particular, we developed and tested models using only the `text` variable, or a combination of the `text` and the `descriptor` variables. We used as baseline a dummy regressor always predicting the mean entropy of the full dataset. We evaluated performance with mean squared error (MSE).

Results on *ANSA* show that the best model improves the MSE by 0.094 (baseline MSE = 0.24), when combining the variable `text` with `descriptor`. Table 18.4 reports cross-validated results for each separate newspaper source, as well as for the full dataset (newspapers + *ANSA*). With the exception of *Il Giornale*, our model always beats the baseline, confirming the feasibility of our approach. Extending the newspaper dataset with the data from *ANSA*, we can observe a general reinforcement of the predicting power of the model, with a range between 0.04 to 0.1 points with respect to the corresponding baseline. The results show that the method is effective in predicting entropy. To verify that entropy is a good proxy for controversy, we manually explored the data which showed that the more entropic news are indeed about controversial topics (for more details see Basile et al. (2018)).

Results for Dutch confirm the validity of the approach outlined for Italian. However, we observed that the average entropy in the Dutch data is already very very small (MSE = 0.0492). This can be a consequence of the smaller amount of data, although it could suggest that controversy in Dutch is a less prominent phenomenon than in Italian.

[4] `https://developers.facebook.com/docs/graph-api`.

Table 18.4 *Cross-validated results for controversy prediction on all datasets*

	BASELINE	STD	MODEL	STD
ilgiornale	**0.21**	0.03	0.22	0.04
ilgiornale+ansa	0.23	0.04	**0.19**	0.03
ilmanifesto	0.15	0.04	**0.11**	0.04
ilmanifesto+ansa	0.24	0.04	**0.14**	0.03
repubblica	0.22	0.07	**0.18**	0.07
repubblica+ansa	0.24	0.04	**0.15**	0.04
corrieredellasera	0.24	0.06	**0.16**	0.06
corrieredellasera+ansa	0.24	0.03	**0.14**	0.04
full_dataset	0.24	0.02	**0.17**	0.03
full_dataset-ansa	0.24	0.03	**0.17**	0.04

18.3.2 Hate Speech Detection in Social Media

We tested the validity of the hate embeddings by applying them in different classifiers and comparing with alternative embedding representations as follows:

- Convolutional Neural Network (CNN), using the implementation of Kim (2014), and experimenting with two different activation strategies: (i) random initialization, by generating word embeddings from the training data itself, i.e., "on-the-fly"; (ii) pre-trained GloVe generic word embeddings (Berardi et al., 2015); (iii) hate embeddings;
- Linear Support Vector Machine (SVM) (Buitinck et al., 2013) with only information from the two different sets of pre-trained embeddings (GloVe generic vs. hate embeddings) by representing each message as a one-dimensional sentence vector with each word replaced with the corresponding embedding average;
- Linear Support Vector Machine (SVM) (Buitinck et al., 2013) integrating the one-dimensional sentence vector representations (GloVe generic vs. hate embeddings) n-gram-based tf-idf model (1–3 word and 2–4 character n-grams).

The models have been trained and tested using the data provided in the context of the EVALITA 2018 task on Hate Speech Detection (`haspeede`) (Bosco et al., 2018). The training/development set comprises 3,000 Facebook comments and 3,000 tweets. We train on 80% of the data (4,800 instances), and test

Table 18.5 *Results for the contribution of different embeddings in CNN and SVM models on the EVALITA training set. Results are reported as averages of 10 different random splits of train and test*

MODEL	CLASS	P	R	F	MACRO F
		EMBEDDINGS ALONE			
CNN on-the-fly embeds	non-H H	0.84 0.77	0.75 0.65	0.79 0.70	0.749
CNN generic embeds	non-H H	0.80 0.74	0.86 0.65	0.83 0.69	0.760
CNN polarized embeds	non-H H	0.82 0.78	0.88 0.68	0.85 0.73	0.786
SVM generic embeds	non-H H	0.77 0.71	0.85 0.60	0.81 0.65	0.728
SVM polarized embeds	non-H H	0.79 0.72	0.84 0.66	0.81 0.69	0.750
		N-GRAMS + EMBEDDINGS			
SVM tf-idf + generic embeds	non-H H	0.84 0.78	0.87 0.74	0.85 0.76	0.806
SVM tf-idf + polarized embeds	non-H H	0.84 0.78	0.86 0.75	0.85 0.76	**0.807**
		N-GRAMS ALONE			
SVM tf-idf	non-H H	0.83 0.78	0.87 0.72	0.85 0.75	0.802

on the remaining 20% (1,200). Performance is assessed using macro F-score reporting the average of ten different runs of random train/test splits. Italian hate embeddings were generated using more than 1 million tokens from fourteen Facebook pages.

The results in Table 18.5 show that despite our embeddings being almost twenty-five times smaller than the generic ones, they yield a substantially better performance both in the CNN model and in the SVM classifier. In the former, they are also more informative than the representations obtained on-the-fly from the training data. In the latter, the contribution of embeddings in general appears rather marginal on top of a more standard SVM model based on n-gram tf-idf information, and the difference is not significant.

With this in mind, we submitted the results of the SVM tf-idf enriched with our polarized embeddings to the actual EVALITA `haspeede` evaluation exercise ranking #6 (out of fourteen submissions) with a macro-F1 of 0.775 (best system macro-F1 0.828; Cimino et al. (2018)) when evaluated against Facebook messages, and #3 (out of fifteen submissions) with a macro-F1 of 0.793 (best system macro-F1 0.799; Cimino et al. (2018)) when evaluated against tweets (Bai et al., 2018b).

We applied the same approach to German (Bai et al., 2018a) by identifying only four relevant Facebook pages, for a total of less than 500,000 tokens for generating the hate embeddings. We submitted a re-trained version of the Italian system to the GermEval 2018 shared task (Wiegand et al., 2018). In this case, the limited amount of data used to generate the hate embeddings (half of the size of Italian) was not enough to obtain results that could fit in the top ten of the fifty-one participating systems (rank #28, macro-F1 of 0.682; best system macro-F1 0.767; Montani and Schüller (2018)), suggesting embedding size is definitely a relevant factor in this approach.

18.3.3 Abusive Language Detection

The Twitter-based polarized embeddings were obtained by selecting keywords that refer to controversial topics, like *abortion, feminism, Black Lives Matter*,[5] as well as politically oriented ones such as *Brexit* or *MAGA*. These keywords have been applied to a large corpus of Twitter messages (Bouma, 2015) in English from 2016. We collected more than 6 millions tweets, for a total of 132,978,104 tokens. We used the GloVe model (Pennington et al., 2014) to generate 200-dimension embeddings, so as to compare their performance against available Glove Twitter embeddings.

We trained separate SVMs using three public datasets, based on Twitter, that are meant to mark abusive language at different levels of granularity, namely SemEval 2019 OffensEval (Zampieri et al., 2019), SemEval 2019 HateEval (Basile et al., 2019), and WH (Waseem and Hovy, 2016), using embeddings as data representation. When evaluated on the corresponding Twitter data for each task, we have observed that our Twitter-based polarized embeddings do not always outperform large generic embeddings (Graumas et al., 2019). This again points toward the importance of size when using this strategy. Additionally, the SVM in which the embeddings are used do not exploit fully the

[5] The list of controversial topics has been developed by manually selecting relevant terms from the Wikipedia list of controversial issues: `https://en.wikipedia.org/wiki/Wikipedia:List_of_controversial_issues`.

Table 18.6 *Out of domain results with Twitter GloVe and polarized embeddings. Best scores in bold*

Trained Model	R	F1 (macro)
OffensEval: GloVe	.30	.47
OffensEval: Polarized	**.32**	**.49**
WH: GloVe	.22	.35
WH: Polarized	**.30**	**.47**
HateEval: GloVe	.17	.29
HateEval: Polarized	**.25**	**.40**

potential of the embedding representations. Lastly, but importantly, as we have pointed out above, manually annotated data, such as the Twitter datasets we used, can carry biases related to their annotations, data selection, and subjectivity of the annotators. These biases might be rather strong and specific and not necessarily well matched by our polarized embeddings.

To address the latter problem associated with bias in manual annotation, we have devised an experiment where we test our models on more ecologically acquired data, rather than manually crafted datasets.

Specifically, we have used the StackOverflow Offensive Comments corpus, a collection of more than 50 million online question and answer messages from the StackOverflow website. We only used the subsets of comments marked as "Rude or Offensive" by the users for a total of 53,978 texts. Comments are flagged by StackOverflow users in a multi-step process and have a near-zero false positive rate (as stated in the accompanying documentation). As a by-product, we also manage to collect evidence on the robustness and portability of the models on out-of-domain data.

Results in Table 18.6 show the performance of the original SVM trained models against the StackOverflow offensive data, comparing generic (GloVe) and polarized embeddings. Precision is omitted as it always corresponds to 1.0 (all messages are offensive because of the extraction method).

In this case, even with a simpler architecture, the polarized embeddings appear to have a stronger predicting power, and thus better portability. Furthermore, we observe different performances according to the different dataset used to trained the SVM, with OffensEval and WH datasets obtaining better results than those using HateEval. This somehow provides indirect evidence about the type of abusive language expressed in the StackOverflow platform, namely offense, sexism, and racism.

18.4 Conclusions

We offered a methodology which is ecologically grounded (important in the subjective context of perspectives in social media), and applied it in different aspects/tasks, showing how to exploit different proxies and their usage case by case. We have also shown that this can be done in different languages and for data coming from different social media platforms, thus addressing the diversity of sources and communities that populate and interact on the Web. We have highlighted advantages and limitations of our strategies. Overall, our methods have proved successful, and can serve as a blueprint for future use, as they can obviously be refined and adapted to other tasks. Our ecologically grounded methodology aims at limiting the introduction of biases, that is, the perspectives of the researchers/scholars, and, at the same time, at directly putting in the foreground the (diversity of) content produced by the users, that is, the perspectives of the Web. In order to further extend this methodology in the study of perspectives, one must decide which aspects of perspective to analyze, define an appropriate task to operationalize them, and find the appropriate cues and strategy to proxy them.

References

Bai, X., Merenda, F., Zaghi, C., Caselli, T., and Nissim, M. 2018a. RuG at GermEval: Detecting offensive speech in German social media. Ruppenhofer, J., Siegel, M., and Wiegand, M. (eds), *Proceedings of the GermEval 2018 Workshop*. Vienna: Austrian Academy of Sciences.

Bai, X., Merenda, F., Zaghi, C., Caselli, T., and Nissim, M. 2018b. RuG@ EVALITA 2018: Hate speech detection in Italian social media. *EVALITA Evaluation of NLP and Speech Tools for Italian*, **12**, 245.

Basile, A., Caselli, T., Merenda, F., and Nissim, M. 2018. Facebook reactions as controversy proxies: Predictive models over Italian news. *Italian Journal of Computational Linguistics*, **4(2)**, 73–90.

Basile, V., Bosco, C., Fersini, E., Nozza, D., Patti, V., Rangel Pardo, F. M., Rosso, P., and Sanguinetti, M. 2019. SemEval-2019 Task 5: Multilingual detection of hate speech against immigrants and women in Twitter. Pages 54–63 of: *Proceedings of the 13th International Workshop on Semantic Evaluation*. Minneapolis, MN: Association for Computational Linguistics.

Berardi, G., Esuli, A., and Marcheggiani, D. 2015. Word embeddings go to Italy: A comparison of models and training datasets. *Proceedings of the 6th Italian Information Retrieval Workshop. Cagliari: CEUR-WS.*

Borra, E., Weltevrede, E., Ciuccarelli, P., Kaltenbrunner, A., Laniado, D., Magni, G., Mauri, M., Rogers, R., and Venturini, T. 2015. Societal controversies in Wikipedia articles. Pages 193–196 of: *Proceedings of the 33rd Annual ACM Conference on Human Factors in Computing Systems*. New York: ACM.

Bosco, C., Dell'Orletta, F., Poletto, F., Sanuguinetti, M., and Tesconi, M. 2018. Overview of the EVALITA hate speech detection (HaSpeeDe) task. Caselli, T., Novielli, N., Patti, V., and Rosso, P. (eds), *Proceedings of the 6th Evaluation Campaign of Natural Language Processing and Speech Tools for Italian (EVALITA '18)*. Turin: CEUR.org.

Bouma, G. 2015. N-gram frequencies for Dutch Twitter data. *Computational Linguistics in the Netherlands Journal*, **5**, 25–36.

Bozdag, E., and van den Hoven, J. 2015. Breaking the filter bubble: Democracy and design. *Ethics and Information Technology*, **17**(4), 249–265.

Bryden, J., Funk, S., and Jansen, V. A. A. 2013. Word usage mirrors community structure in the online social network Twitter. *EPJ Data Science*, **2**(1), 3.

Buitinck, L., Louppe, G., Blondel, M., Pedregosa, F., Mueller, A., Grisel, O., Niculae, V., Prettenhofer, P., Gramfort, A., Grobler, J., Layton, R., VanderPlas, J., Joly, A., Holt, B., and Varoquaux, G. 2013. API design for machine learning software: Experiences from the scikit-learn project. Pages 108–122 of: *ECML PKDD Workshop: Languages for Data Mining and Machine Learning*. Prague.

Cimino, A., De Mattei, L., and Dell'Orletta, F. 2018. Multi-task learning in deep neural networks at EVALITA 2018. Pages 86–95 of: *Proceedings of the 6th Evaluation Campaign of Natural Language Processing and Speech Tools for Italian*. EVALITA '18, Turin: CEUR-WS.

Coletto, M., Garimella, K., Gionis, A., and Lucchese, C. 2017. Automatic controversy detection in social media: A content-independent motif-based approach. *Online Social Networks and Media*, **3**, 22–31.

Derczynski, L., Bontcheva, K., Liakata, M., Procter, R., Wong Sak Hoi, G., and Zubiaga, A. 2017. SemEval-2017 Task 8: RumourEval: Determining rumour veracity and support for rumours. Pages 69–76 of: *Proceedings of the 11th International Workshop on Semantic Evaluation*. SemEval '17. Vancouver: Association for Computational Linguistics.

Dori-Hacohen, S., and Allan, J. 2015. Automated controversy detection on the Web. Pages 423–434 of: *European Conference on Information Retrieval*. Vienna: Springer.

Exner, P., Klang, M., and Nugues, P. 2015. A distant supervision approach to semantic role labeling. Pages 239–248 of: *Proceedings of the Fourth Joint Conference on Lexical and Computational Semantics*. Denver: Association for Computational Linguistics.

Founta, A.-M., Djouvas, C., Chatzakou, D., Leontiadis, I., Blackburn, J., Stringhini, G., Vakali, A., Sirivianos, M., and Kourtellis, N. 2018. Large scale crowdsourcing and characterization of Twitter abusive behavior. Pages 491–500 of: *Proceedings of the Twelfth International Conference on Web and Social Media*. ICWSM '18. Stanford, CA.

Garimella, K., De Francisci Morales, G., Gionis, A., and Mathioudakis, M. 2016. Quantifying controversy in social media. Pages 33–42 of: *Proceedings of the Ninth ACM International Conference on Web Search and Data Mining*. New York: ACM.

Geva, M., Goldberg, Y., and Berant, J. 2019. Are we modeling the task or the annotator? An investigation of annotator bias in natural language understanding datasets. *arXiv preprint arXiv:1908.07898*.

Go, A., Bhayani, R., and Huang, L. 2009. Twitter sentiment classification using distant supervision. *CS224N Project Report, Stanford*, **1**(12).

Gorrell, G., Kochkina, E., Liakata, M., Aker, A., Zubiaga, A., Bontcheva, K., and Derczynski, L. 2019. SemEval-2019 Task 7: RumourEval, determining rumour veracity and support for rumours. Pages 845–854 of: *Proceedings of the 13th International Workshop on Semantic Evaluation*. Minneapolis: Association for Computational Linguistics.

Graumas, L., David, R., and Caselli, T. 2019. Twitter-based polarised embeddings for abusive language detection. Pages 1–7 of: *2019 8th International Conference on Affective Computing and Intelligent Interaction Workshops and Demos*.

Gururangan, S., Swayamdipta, S., Levy, O., Schwartz, R., Bowman, S. R., and Smith, N. A. 2018. Annotation artifacts in natural language inference data. Pages 107-112 of: *Proceedings of the 2018 Conference of the North American Chapter of the Association for Computational Linguistics: Human Language Technologies, Vol. 2*. New Orleans: Association for Computational Linguistics.

Hanselowski, A., Avinesh P. V. S., Schiller, B., Caspelherr, F., Chaudhuri, D., Meyer, C. M., and Gurevych, I. 2018. A retrospective analysis of the fake news challenge stance-detection task. Pages 1859–1874 of: *Proceedings of the 27th International Conference on Computational Linguistics*. Santa Fe: Association for Computational Linguistics.

Kennedy, G., McCollough, A., Dixon, E., Bastidas, A., Ryan, J., Loo, C., and Sahay, S. 2017. Technology solutions to combat online harassment. Pages 73–77 of: *Proceedings of the First Workshop on Abusive Language Online*.

Kim, Y. 2014. Convolutional neural networks for sentence classification. Pages 1746–1751 of: *Proceedings of the 2014 Conference on Empirical Methods in Natural Language Processing*. Doha: EMNLP.

Mejova, Y., Zhang, A. X., Diakopoulos, N., and Castillo, C. 2014. Controversy and sentiment in online news. *arXiv preprint arXiv:1409.8152*.

Mikolov, T., Chen, K., Corrado, G., and Dean, J. 2013. Efficient estimation of word representations in vector space. *arXiv preprint arXiv:1301.3781*.

Mintz, M., Bills, S., Snow, R., and Jurafsky, D. 2009. Distant supervision for relation extraction without labeled data. Pages 1003–1011 of: *Proceedings of the Joint Conference of the 47th Annual Meeting of the ACL and the 4th International Joint Conference on Natural Language Processing of the AFNLP: Vol. 2*. Suntec: Association for Computational Linguistics.

Montani, J. P., and Schüller, P. 2018. TUWienKBS at GermEval 2018: German abusive tweet detection. Page 45 of: *14th Conference on Natural Language Processing KONVENS*, Vol. 2018.

Nobata, C., Tetreault, J., Thomas, A., Mehdad, Y., and Chang, Y. 2016. Abusive language detection in online user content. Pages 145–153 of: *Proceedings of the 25th International Conference on World Wide Web*. Republic and Canton of Geneva: International World Wide Web Conferences Steering Committee.

Pariser, E. 2011. *The filter bubble: What the Internet is hiding from you*. Penguin.

Paun, S., Carpenter, B., Chamberlain, J., Hovy, D., Kruschwitz, U., and Poesio, M. 2018. Comparing Bayesian models of annotation. *Transactions of the Association for Computational Linguistics*, **6**, 571–585.

Pennington, J., Socher, R., and Manning, C. D. 2014. GloVe: Global vectors for word representation. Pages 1532–1543 of: *Empirical Methods in Natural Language Processing*.

Pool, C., and Nissim, M. 2016. Distant supervision for emotion detection using Facebook reactions. Pages 30–39 of: *Proceedings of the Workshop on Computational Modeling of People's Opinions, Personality, and Emotions in Social Media (PEOPLES)*. Osaka: COLING 2016.

Schmidt, A., and Wiegand, M. 2017. A survey on hate speech detection using natural language processing. Pages 1–10 of: *Proceedings of the Fifth International Workshop on Natural Language Processing for Social Media*.

Seargeant, P., and Tagg, C. 2018. Social media and the future of open debate: A user-oriented approach to Facebook's filter bubble conundrum. *Discourse, Context & Media*, **27**, 41–48.

Timmermans, B., Aroyo, L., Kuhn, T., Beelen, K., Kanoulas, E., van de Velde, B., and van Eerten, G. 2017. ControCurator: Understanding controversy using collective intelligence. *Collective Intelligence 2017*. Sage/ACM.

Tsuchiya, M. 2018. Performance impact caused by hidden bias of training data for recognizing textual entailment. *Proceedings of the Eleventh International Conference on Language Resources and Evaluation*. LREC '18. Miyazaki: European Languages Resources Association.

van Aken, B., Risch, J., Krestel, R., and Löser, A. 2018. Challenges for toxic comment classification: An in-depth error analysis. Pages 33–42 of: *Proceedings of the 2nd Workshop on Abusive Language Online (ALW2)*. Brussels: Association for Computational Linguistics.

Waseem, Z., and Hovy, D. 2016. Hateful symbols or hateful people? Predictive features for hate speech detection on Twitter. Pages 88–93 of: *Proceedings of the NAACL Student Research Workshop*. San Diego, CA: Association for Computational Linguistics.

Wiegand, M., Ruppenhofer, J., and Kleinbauer, T. 2019. Detection of abusive language: The problem of biased datasets. Pages 602–608 of: *Proceedings of the 2019 Conference of the North American Chapter of the Association for Computational Linguistics: Human Language Technologies, Vol. 1: Long and Short Papers*. Minneapolis, MN: Association for Computational Linguistics.

Wiegand, M., Siegel, M., and Ruppenhofer, J. 2018. Overview of the GermEval 2018 shared task on the identification of offensive language. Pages of 1–10: *Proceedings of GermEval 2018, 14th Conference on Natural Language Processing*. KONVENS 2018. Vienna: Austrian Academy of Sciences.

Wulczyn, E., Thain, N., and Dixon, L. 2016. *Ex machina: Personal attacks seen at scale*. Pages of: Proceedings of the 26th International Conference on World Wide Web. Republic and Canton of Geneva.

Zampieri, M., Malmasi, S., Nakov, P., Rosenthal, S., Farra, N., and Kumar, R. 2019. Predicting the type and target of offensive posts in social media. In: *Proceedings of the 2019 Conference of the North American Chapter of the Association for Computational Linguistics: Human Language Technologies. Vol. 1*. Minneapolis: Association for Computational Linguistics.

19

GRaSP: A Model for the Perspective Web

Piek Vossen and Antske Fokkens

19.1 Introduction

This book has opened a Pandora's box about the ways in which we communicate online and the impact it can have. We have also seen a wide parade of concepts and models from the different disciplines to describe and explain this phenomenon, with various proposals for solutions or improvements for current practices.

Finding knowledge and information on the Web is becoming an almost trivial task, and understanding its value, the credibility, and perspective of its source is becoming increasingly problematic. While the Internet and other digital modes of communication have made access to knowledge and information much easier, the burden on the users of the Web to filter and judge information has increased. Technology lags behind when it comes to providing users with proper tools to track information provenance, the perspective of the source, and the validity of information. It is a common saying that on the Web you can find any truth that is convenient, yet we have little knowledge about how beliefs actually spread and how this impacts our view on the world. Many questions with respect to the impact of being online will remain unanswered unless we come to grips with these social communication processes. When do online discussions in distributive communities lead to convergence and when to polarization? How does social filtering work and does it change as a result of online debates? What is the impact on identity and group membership when dealing with convergent or hostile beliefs?

Besides the fact that, nowadays, we can track many social activities on the Web and analyze what people "talk about", the language that people use is in itself a source of information about the attitudes and perspectives that people hold. There is no such thing as simple text. We often express our emotions, opinions, and positions verbally in subtle ways that can sometimes be made explicit more easily by computer software than by human observers. Automatic mining of subjectivity and perspective relations from language could be exploited to learn more about the impact of communication on the Web. The availability of digital data at all these different levels opens up the unique opportunity to dynamically model information, knowledge, and mediated communication as well as their social implications. This can be done from the lowest micro-level of symbolic data (such as texts) to interpretation and knowledge, up to the higher macro-levels of (1) social groups and identity, (2) social activity and dynamics, and (3) societal impact.

One of the challenges involved in dealing with the wealth of information provided on the World Wide Web (WWW) today is determining how reliable and trustworthy information is. It can be difficult to determine what the source of specific information is. Even if the source is known it can still be difficult to determine how reliable information is. Reliability and trust can, in principle, not be measured by an objective score, and it will, in the end, be up to the users to decide whether to believe the information provided or to follow an alternative or even opposite point of view. To support this process and help users find reliable information, we believe it is possible and desirable to provide as much insight into the information as possible. We envision a Perspective Web as an enriched version of the WWW that augments the Web's content with information that increases transparency in provenance of information, and provides access to information from groups with alternative viewpoints together with as much information as possible about the source of information. The kind of information we would like to provide would answer questions such as: What expertise does a person or team have? What (conflict of) interest? Are claims supported by independent observations or scientific research?

This closing chapter addresses the question of how such information can be modeled formally in a way that can be used in automatic search as an implementation of **the Perspective Web**. In particular, we illustrate how the Grounded Representation of Source Perspective framework (henceforth, GRaSP) (Fokkens et al., 2017b; Son et al., 2016; Vossen and Fokkens, 2021) provides a generic model for representing information related to perspectives.

In this chapter, we focus on information that is directly related to other chapters in this book. We look at phenomena users of the Web would benefit from, information researchers need to study to learn how perspectives are presented or spread on the Internet, and how to represent the information that we can currently extract automatically, including information on the reliability of the tools. We are aware that not all information that is mentioned in the chapters and our examples is actually available, but this part of the challenge lies out of the scope of this chapter.

The remainder of this chapter is structured as follows. We first provide a high-level representation for GRaSP as a model for the Perspective Web. In Section 19.3, we describe the requirements for this model and the details for the model itself. We describe related models in Section 19.4. Finally, we conclude and describe a future perspective on the road map to the Perspective Web in Section 19.5.

19.2 A Model for the Perspective Web

Starting from the basic model of communication introduced by Shannon (1948) and discussed in Parts II and IV of this book, we can explain GRaSP as a model for capturing communication processes. GRaSP, in its most basic form, defines relations between (1) the source and a signal, (2) segments of the signal and interpretations of what the signal refers to (what is the source talking about), and (3) segments of the signal and interpretations of what viewpoints the signal expresses (e.g., negating vs. confirming, associated sentiment, emotions). The perspective of a source is then expressed by the full range of interpretations of the signals they transmitted. In other words, a perspective is made up of what a source chooses to talk about or show (the relations expressed in 2) and how the source talks about this (the relations expressed in 3).

In our communication, the signal is often a text that contains a statement referencing things in the world and expressions that make clear what the perspective is of a source to the statements. Consider the next tweet from Donald Trump:

> I'm not against vaccinations for your children, I'm against them in 1 massive dose.Spread them out over a period of time & autism will drop! – Donald J. Trump (@realDonaldTrump) September 4, 2014
> www.businessinsider.nl/trump-vaccines-autism-wrong-2017-1/?international=true&r=UK

Trump makes a few statements in this tweet: *vaccination of children, vaccination causes autism*. Furthermore, he expresses a few perspectives to these statements: he is in favor of vaccinating children, vaccination in massive doses causes autism, vaccination spread over time does not cause autism. GRaSP formally models where in the text the statements can be found, what these statements are semantically, where the perspective is expressed, whose perspective it is, and when it was made. If we want to identify these relations in large volumes of data, they need to be detected using Natural Language Processing technology. If this is done correctly, it is possible to store the result

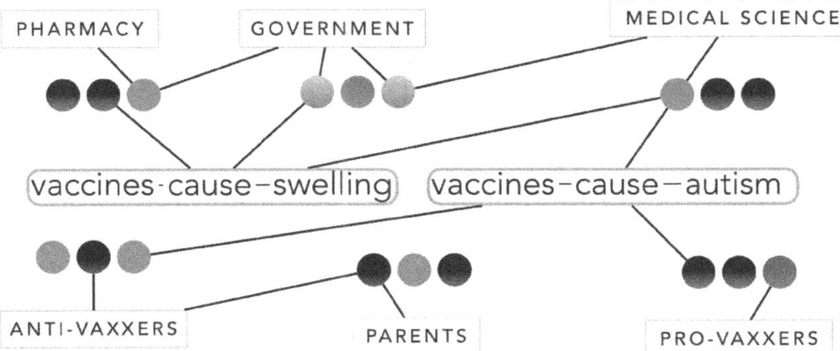

Figure 19.1 A high-level example of a Perspective Graph generated in GRaSP that represents statements as triple relations and groups various source perspectives on shared statements. Each circle represents a unique set of perspective values expressed and shared by a group of sources. Perspectives can take different values such as confirm, denial, certainty, sentiment, judgment, and so on. Social groups can be defined on the basis of shared expressions and related to other metadata such as their stance (pro or anti) or their interest (parents, government, pharmacy, science)

in the GRaSP model and search for all cases in which the same statement is made, who made it, when was it made, and who is the source.

Figure 19.1 shows an schematic example of what we call a Perspective Graph that can be extracted from data modeled in GRaSP. It represents two statements as triple relations: vaccines cause swelling and vaccines cause autism. It further groups different source perspectives on shared statements. Each colored circle is a unique perspective expressed by a source at a specific moment in time. Perspectives can take different values such as confirm, denial, certainty, sentiment, judgment, which is reflected by the colors. Social groups can be defined on the basis of shared expressions and related to other metadata such as their stance (pro or anti) or their interest (parents, government, pharmacy, science). Using such a modeling of the data, systems can trace the dynamics of perspectives in time, over regions, and different social groups.

Perspective Graphs such as in Figure 19.1 can be automatically extracted from the current Web using advanced NLP technologies that can interpret posted opinions and information with the complex and rich social contexts of digital communication and storing the output according to the GRaSP model.

19.3 The GRaSP Model

This section decribes the GRaSP model largely following our earlier work (Fokkens et al., 2017b; Son et al., 2016; Vossen and Fokkens, 2021). We first outline the requirements of the model, starting with general requirements augmented with additional requirements for building a Perspective Web based on various themes from individual chapters. We then provide a brief introduction to the principles of the Semantic Web before we describe the model itself.

19.3.1 Requirements of the Model

In Fokkens et al. (2017b), we defined six general requirements for the GRaSP model:

1. Represent various perspectives on the same entity, proposition, or topic next to each other.
2. Represent the source of each perspective, so that users can, for example, select all perspectives of a specific source; group sources according to shared or conflicting views on a given content; find all sources that have a perspective on the same content or share a perspective; and find available background information about the source.
3. Semantically compare the (propositional) content across statements and represent whether sources mention the same, similar, or related content (e.g., more or less specific), or a different framing of content (e.g., "murdered," which is intentional, or "killed" which may be accidental).
4. Represent a wide range of perspective-related phenomena, including sentiment, emotion, judgment, negation, certainty, speculation, reporting, framing, and salience.
5. Make alternative interpretations of the same statement explicit, since statements might be (deliberately) ambiguous, not well formulated, or difficult to process with Natural Language Processing (NLP).
6. Provide the full provenance of any information in GRaSP. Next to the source, it should provide information about how this perspective was analyzed (e.g., expert analysis of a text, crowd annotations, text mining).

The first three requirements allow users to place various perspectives next to each other so that they can compare, among others, which sources agree or disagree on what, which sources change their mind, which sources speculate and whether their predictions were accurate. In addition, they would allow identification of all content and stances given on a specific topic by a source and, for example, display this on a timeline. The fourth and fifth requirements ensure that the model is flexible enough to support various needs of end-users as well as to accommodate the variation of information provided by different systems or datasets.

We connect the above general requirements to the different chapters of this book, starting with Part II. The three levels (micro Chapter 5, meso Chapter 6, and macro Chapter 7) introduced in **Social Impact** provide a nice illustration of what is needed to look at details involved around perspectives (micro level), what needs to be modeled from a social point of view (meso level), and what is needed to study the big picture (macro level). The requirements from the micro perspective are directly covered in the original design and examples of GRaSP: they confirm the need to capture linguistic details at a fine-grained level and, in particular, the need to explicitly model what is implied by the actual text. The meso-level addresses the importance of social groups in determining beliefs: the tendency of people to trust information that comes from like-minded people. In other words, when they agree with someone or a group on one point of view, they are also more likely to agree with them on other topics even if these topics are not related (at all). This adds the requirement that GRaSP must be able to provide insight into social interaction and social networks. Finally, the macro - level deals with (the myth of) echo chambers and the need to present diverse points of view and in particular those of minorities. This level brings the micro and meso levels together: it requires precise representation of the information or point of view that is expressed together with information of the social context of the source (is it diverse? Mainstream or minority?).

Part III of the book on **mediating perspectives** introduces the following additional requirements. In Chapter 9, Hoyt explains that metadata plays a vital role in determining which pages are presented to users in search and warns that metadata is not always reliable. It is often provided by the owners of a site with specific interest in goals in mind. For researchers as well as critical users of the Web, information about metadata and who created it (the site's owners or another independent source) is vital and GRaSP should be able to introduce this. Chapter 11 by Neumayer illustrates the importance of the interaction between the content (what information is provided), the form (how is this provided), and reception (how is it received). The first two components are covered by the original requirements of GRaSP, though Neumayer explicitly adds the purpose behind the form in which information is presented (e.g., to draw attention).

Chapter 12 investigates various metrics that can be used for assessing the quality of information that is provided. Although the chapter discusses assessments mainly by the crowd and experts, many of their quality dimensions can also be derived from data modeled in GRaSP. **Accuracy** relates to the veracity of claimed statements, which GRaSP can provide as a score based on the degree there are opposing perspectives, for example, nobody denies that children are being vaccinated. **Precision** of the statement can be provided by the semantics of the statement itself, that is, how many triple relations are provided about a topic, how nuanced are the values for properties (e.g., precise number or global numbers, precise locations, specific names). Similarly, **Completeness** can be weighed by the density and richness of the statements. Finally, **Trustworthiness** can be derived by the transparency on the sources, the explicitness of the source references in the texts, and the degree sources are consistent. Obviously, metadata on the authority of sources that is not in the text itself needs to be derived elsewhere.

The information that is extracted by the methods in Part IV on **Mining and Modeling** perspectives is mostly covered by the original requirements and the requirements added above. Chapter 14 gives a comprehensive overview of the state of the art in NLP for detecting sources, perspectives, and propositions in texts, which make up the main components in GRaSP. In Chapter 15, Chen et al. consider NLP as a way of aggregating a comprehensive and less-biased overview of perspectives, where Verberne in Chapter 17 addresses personalized search and the need to generate explanations. Similarly, Caselli and Nissim in Chapter 18 harvest perspectives from social media with a similar purpose. In all these cases, GRaSP can play a two-fold role: it can provide machine-readable representations of information that may be used for personalized search and to represent what information is used in order to come to a recommendation (both for personalization as in general).

19.3.2 The Semantic Web

It follows from the above that a central point in providing full insight in to a topic lies in connecting pieces of information in a more transparent and explicit way. We therefore represent information as Linked Data following Semantic Web standards. Linked Data represents information in RDF (Resource Description Framework),[1] an extensible knowledge representation model specifically designed to connect information and reason over it. The RDF schema presents information in triples consisting of a subject, predicate, and object. Entities, classes, and relations are represented by unique identifiers called internationalized resource identifiers (IRIs).[2] We provide a simple example to explain the principle. Consider the following triples from DBpedia,[3] a database representing information from Wikipedia as Linked Data:

1 www.w3.org/RDF/.

2 RDF versions prior to RDF 1.1 used uniform resource identifiers. The term URI is therefore also commonly used to indicate identifiers in RDF.

3 https://wiki.dbpedia.org.

```
http://dbpedia.org/resource/Amsterdam
http://dbpedia.org/ontology/isPartOf
http://dbpedia.org/page/North_Holland

http://dbpedia.org/page/North_Holland
http://dbpedia.org/ontology/isPartOf
http://dbpedia.org/page/Netherlands
```

These triples state that Amsterdam is part of North Holland and that North Holland is part of the Netherlands. Because we can define being part of something as a transitive relation, we can infer that Amsterdam is part of the Netherlands. Now consider two more triples that are found in two different databases:

```
http://dbpedia.org/resource/Amsterdam
http://historic_ontology.org/cityRightDate
"1300"^^xs:date.

http://dbpedia.org/resource/Amsterdam
http://historic_ontology.org/cityRightDate
"1306"^^xs:date.
```

These statements were created independently and make use of the same (also made-up) generic ontology that includes the relation of the date on which a place obtained city rights, where one source places this in the year 1300 and the other in 1306. The value is a literal, specified as a date. Because the IRI for Amsterdam is also shared in both statements, we can compare the actual content of the statements. By linking information from all three resources, we now have information about the location of Amsterdam (being part of North Holland and the Netherlands) as well as when it received city rights. Amsterdam only received these rights once, but we have two conflicting values for this.

The Semantic Web has no problem with conflicting information standing next to each other, though, naturally, researchers have tried to find solutions to help users determine the reliability of information. An important contribution in this aspect came from PROV-O (Lebo et al., 2013), an ontology for modeling the provenance of information. With GRaSP, we build on top of this basic information on provenance and introduce a framework that is targeted toward citing information and pointing back to textual sources. In addition, GRaSP provides the means to indicate which level of uncertainty a source attributes to a statement and whether they express a specific sentiment or emotion.

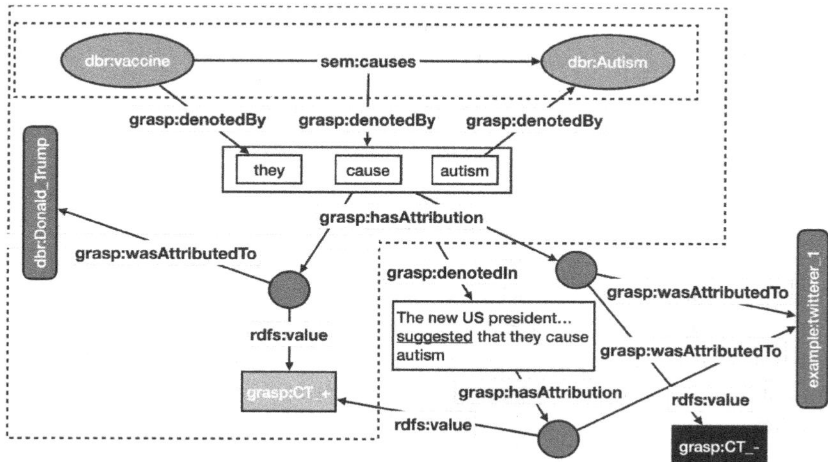

Figure 19.2 Schematic representation of GRaSP of a tweet on vaccination from an anonymous twitterer

19.3.3 Applying GRaSP to a Tweet

Consider the following tweet from an anonymous twitterer:

[The new US president <SOURCE>] has said [he <SOURCE>] believes that [vaccines are harmful <CLAIM>] and has [repeatedly and erroneously <NEGATIVE>] suggested that [they cause autism <CLAIM>]

Figure 19.2 illustrates how GRaSP represents the perspectives on *vaccines cause autism* as expressed in this tweet.

The statement that vaccines cause autism is represented at the top of the image. We typically use the Simple Event Model (SEM, van Hage et al. (2011)) that is designed to model events and states as a generic framework for statements. We use DBpedia IRIs to denote "vaccine" and "autism" and they are related to each other through the causal relation defined in SEM. The statement is linked to a component of the aforementioned tweet using the GRaSP relation denotedBy.[4]

Next, we model the perspective on the statement expressed by a specific source in this text. Recall that according to the tweet, Trump asserted that vaccines cause autism and that the twitterer qualified this statement as "erroneous." In this case, we have thus two separate attributions linked to the same statement, indicated by the leftmost and rightmost circles. The left circle represents Trump

[4] The tweet itself, as well as its components, can also be represented by IRIs, which then point to the exact text that expresses a statement and can be resolved through the Twitter API. This is not done here for reasons of simplification.

confirming it (without expressing doubt), indicated by value CT_+. The right circle represents the twitterer confidently denying it (value CT_-). Finally, Trump's view is expressed by a secondary source in this case. Figure 19.2 illustrates that we can link the entire representation of Trump asserting that vaccines cause autism (without expressing doubt) as represented in the tweet and that the twitterer is the source of Trump's view. If we find Trump's original tweets on the subject, we can directly compare their content to the reported content in this example and thus automatically verify whether his view is reported correctly.

Figure 19.2 merely provides a simplified and partial illustration of what can be achieved by GRaSP. For the example above, we can also make explicit that the twitterer's opposed view is expressed erroneously, that vaccines causing autism is a more specific statement supporting the view that they are harmful and so on. More importantly, though, because we also use IRIs to denote the source, we can provide information about this source: what other statements does this source make? What is known about their background? This is a vital component in determining how reliable information is and how it should be interpreted. Furthermore, GRaSP can be used to model alternative interpretations depending on concepts (following Fokkens et al. (2016) for representing concept drift) or for identifying patterns that result in stereotyping (representing information identified in Herbelot et al.'s (2012) or Fokkens et al.'s (2018) approaches).

This ability to connect statements with all their nuances, background information, and sources makes GRaSP an extremely powerful framework for modeling perspectives. This leaves us with the challenge of how to obtain these detailed representations from large amounts of text.

19.4 Related Work

The GRaSP framework is a flexible, generic framework for representing perspectives. It provides possibilities of indicating the source of information, specific (and possibly alternative) interpretations of a statement, information about the context of the statement, and values attributed to the statement (either through context or to an alternative source). We discuss two types of related work: (1) the most closely related models that also cover one of the aforementioned aspects and (2) models that are frequently used in combination with GRaSP following Fokkens et al. (2017b).

The Web Annotation Data Model (OA)[5] of the W3C is closely related to GRaSP in that it is designed to represent associations between different pieces of information, such as comments on shared images, articles, or videos. This

[5] www.w3.org/TR/annotation-model/.

generic model represents annotations as a directed relation from a body which is "related to and typically about"[6] a target (e.g., a relation from the comment to the image it is about). In principle, the specification of sources and interpretations that we represent in GRaSP can also be represented as annotations, but there are three differences that make it less suitable for the goals set by GRaSP. First, the directionality of the relation between body and target in OA, which indicates that the body is seen as an enrichment of the target. This vision makes sense when making information such as the factuality or sentiment expressed in a statement explicit. It is, however, not compatible with the idea that combined mentions of an instance also collectively determine its semantics. The neutral representation by GRaSP is therefore more suitable for research around stereotyping through language use (Fokkens et al., 2018; Herbelot et al., 2012) or conceptual change (Fokkens et al., 2016). Second, the OA specification states that targets are *dereferencable*. GRaSP aims to also provide information about resources provided by external resources, behind paywalls, and/or restricted by licenses for which dereferenceability cannot be guaranteed.

Another model that is closely related to GRaSP is Marl,[7] which can represent subjective opinions in text. It shares GRaSP's flexibility of being compatible with other models to cover various aspects of opinions. The Onyx ontology[8] uses Marl for representing emotions and made use of Lemon (Buitelaar et al. 2013)[9] for representing lexical semantic information. GRaSP differs from Marl in that GRaSP is not restricted to text. In fact, any multimodal signal can be interpreted and modeled. In recent work, GRaSP has been used to model Human–Robot communication in physical enviroments, relating multimodal signals and communication to interpretations of situations (Vossen et al., 2018, 2019a, 2019b, 2019c). GRaSP furthermore carefully separates information that is confounded in Marl: Marl uses one central node that refers to an object/feature (instance in GRaSP) and the literal text that reflects the opinion (mentions). As such, Marl merely provides a link between an opinion and the text expressing it without making the opinion holder explicit. It can therefore not be used to collect all perspectives provided by a specific source or to perform network analysis on those sources sharing an opinion. GRaSP can be used to model the information needed for such analyses. GRaSP's separation of these layers makes it more flexible in dealing with alternative interpretations of mentions, at both the attribution and instance layer. Finally, GRaSP is not limited to explicitly subjective opinions but can connect all stances taken by a source (including factual statements).

6 www.w3.org/TR/annotation-model/, Section 2.
7 http://gsi.dit.upm.es/ontologies/marl/.
8 www.gsi.dit.upm.es/ontologies/onyx/.
9 http://lemon-model.net.

Nanopublications[10] (Groth et al., 2010) is a framework to address scientific claims similar to our modeling in GRaSP. A nanopublication consists of an assertion, a provenance component, and publication information. The assertion is an atomic unit of information similar to a statement in GRaSP. The provenance layer relates the assertion to the source, for example, a specific scientific study or method, where the publication corresponds to the bibliographical reference. Nanopublications do not consider the text as a linguistic structure and cannot incorporate NLP processing (e.g., alternative interpretations, associations related to choice of words, etc.).

GRaSP can be combined with various existing models. The NLP Interchange Format (NIF, Hellmann et al. 2013) is an RDF/OWL vocabulary for representing NLP annotations in a common way to foster interoperability between NLP tools, language resources, and annotations. The core of NIF consists of a vocabulary and a URI design that permit describing strings and substrings, to which arbitrary annotations can be attached using vocabularies external to NIF. NIF itself does not specifically address the representation of source or attribution information, but can be combined with GRaSP. GRaSP bases the format of IRIs of mentions on NIF and uses it to represent some mention layer attributes (e.g., char offset in the text). By combining GRaSP with Bouquet et al.'s (2005) concept representation and Lemon as done in Fokkens et al. (2016), GRaSP can be used to model complex phenomena such as concept drift. GRaSP uses the grounding relations provided by GAF (Fokkens et al., 2013). GRaSP's main contribution compared to GAF is that GRaSP adds an attribution layer tying sources and their stances to mentions. Similar to Nanopublications, we use PROV (Moreau et al., 2013) to model the provenance of mentions and interpretations made on them (i.e., to model the NLP process following Ockeloen et al. (2013).

This flexibility of GRaSP is what makes it a suitable model to support a Perspective Web: it can be used to represent nuances expressed by natural language in fine-grained manner, as well as generic patterns about sources and networks. Representing provenance on top of this is essential for the transparency required of a Perspective Web.

19.5 A Future Perspective on the Web

Earlier in this book, we discussed the issue that information and knowledge in online debates are dispersed and difficult to grasp, as many different sources post information discontinuously and do not address each other's claims. In this chapter, we described the Semantic Web model GRaSP, which allows

[10] http://nanopub.org/wordpress/.

reasoning over statements, claims, perspectives, and sources on a large scale. We showed how the GRaSP framework can be used to connect claims that express the same or different perspectives on a statement and that it provides the means to find the original source of a claim, verify the claim itself, or find background information on the person or organization that made the claim. The semantic modeling of the propositional relations makes it possible to find sources making related claims, and to contrast perspectives across individual sources and groups. We related the model to the broad interdisciplinary discussion of digital media communication, as explained in the other parts of the book.

The Perspective Web is a projection on the current Web that shows the perspective of the sources on the information and knowledge that we find. The Perspective Web makes explicit and transparent where information comes from, how it changed, who is behind it, what the different positions are. It allows for eliciting the dynamics of debates, opinions, and views on a global scale. The final question that remains is: **can we build the Perspective Web**?

GRaSP can capture the output of NLP technologies that process texts on a massive scale to find attributions and claims. The former projects BiographyNet (Fokkens et al., 2017a) and Newsreader (Vossen et al., 2016), which also laid the foundation for GRaSP, created complex NLP pipelines for so-called reading machines. These reading machines detect the *who*, *what*, *where*, and *when* from any text, as well as sentiments, judgments, and attribution values. Figure 19.3 shows a range of NLP modules that each output a specific interpretation for a single sentence. Each module is trained and or tested on a specific set of texts annotated by people. State-of-the-art test results of some of the most essential models for requiring such information range from 80–90% accuracy. By annotating more texts for specific topics and genres, these modules can be tuned for specific purposes, such as the output for the Perspective Web. Without fine-tuning the performance of these models can drop by 20% or more. Reading machines have been developed for different languages, generating output that is semantically interoperable.

The output of these NLP modules can be represented in GRaSP. Likewise, the information expressed in different languages and from many different sources can be combined in one interoperable model. In BiographyNet and NewsReader, GRaSP was used to store the NLP output for hundreds of thousands of biographies and millions of news articles in the form of billions of triples. Currently, GRaSP is used in various other projects, and we hope that future interdisciplinary research will show how GRaSP can be used for

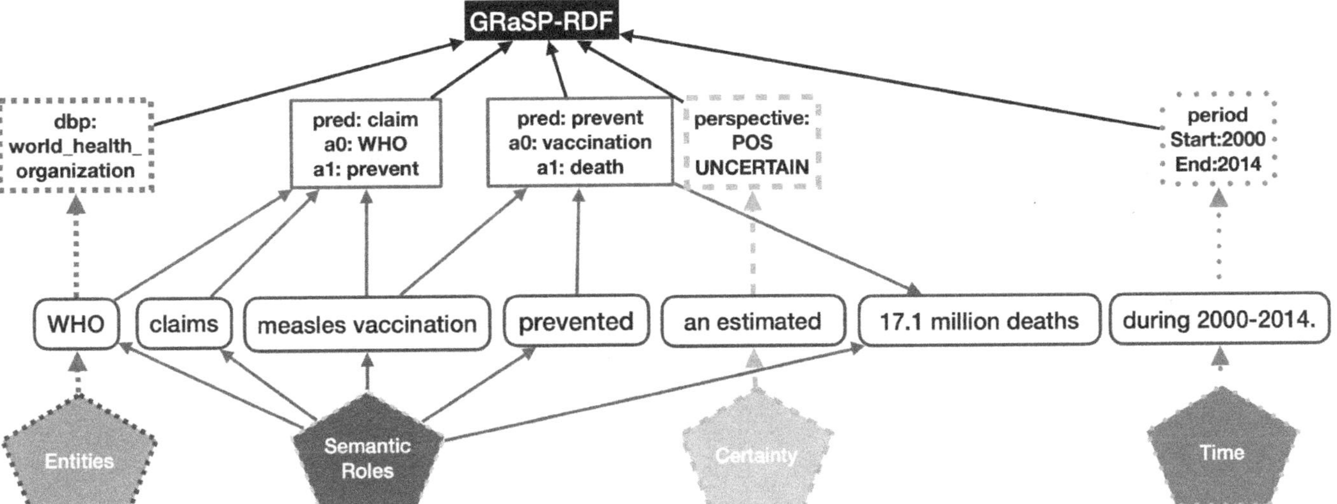

Figure 19.3 NLP modules, representing a reading machine for extracting claims and perspectives from a single text sentence

sociological and media analysis studies on complex and large-scale digital communication and become a cornerstone for building a Perspective Web in the future.

The NLP modules used in these projects are currently being replaced by the latest Deep Learning modules based on transformer-based language models (Devlin et al., 2018; Peters et al., 2018). These models have revolutionized the performance of NLP, especially when fine-tuned for specific tasks and types of text. But will it be enough to build the Perspective Web? This is still to be seen. For now, there are also new hurdles introduced by these models, such as biases that cannot easily be removed and can have a devastating effect on perspectives extracted from text, for example, gender biases in language models which overpredict negative sentiment on women which need to be mitigated (Costa-jussà et al., 2019).

We foresee three important developments in the near future that will be decisive for the development of the Perspective Web: (1) more robust and adaptive models that can deal with the large variation of language and communication, (2) NLP models that take the large social context of communication into account when pieces of text are represented, and (3) more in-depth research on subtle forms of stereotyping picked up by language models. With respect to point (2), we believe that GRaSP is in a good position as it represents such contextual social relations. These relations, when built up by processing sufficient data in some social context such as, for example, a debate on vaccination, can be mixed in with the representation of a single piece of text. This will result in a text representation that is not only the result of rich language models such as ELMO (Peters et al., 2018) or BERT (Devlin et al., 2018) but also from the GRaSP embedding. Obviously, we need to annotate a lot more data with these contexts to be able to test such more comprehensive approaches. A start has been made for the vaccination debate by Morante et al. (2020). With respect to point (3), distributional semantics plays a two-fold role. On the one hand, NLP methods are hindered by biases that language models take over from the examples they are created from. On the other hand, distributional methods can be used to study the extent to which available texts express stereotypes (Fokkens et al., 2018; Herbelot et al., 2012) and as such potentially contribute to the solution.

Finally, who will build the Perspective Web? In the current situation, it is up to the tech companies to read this book and provide the infrastructure to turn the Web into a more transparent platform. They need to implement the technology that supports and if necessary intervenes in the communication society in the future. We, the users of the Web, can only try to make our demands heard so that they take their responsibility.

References

Bouquet, P., Serafini, L., and Stoermer, H. 2005. Introducing context into RDF knowledge bases. Pages 14–16 of: *Proceedings of the 2nd Italian Semantic Web Workshop (SWAP 2005)*. Trento, Italy: CEUR-WS.org

Buitelaar, P., Arcan, M., Iglesias, C. A., Sánchez-Rada, J. F., and Strapparava, C. 2013. Linguistic linked data for sentiment analysis. Page 1 of: *2nd Workshop on Linked Data in Linguistics*.

Costa-jussà, M. R., Hardmeier, C., Radford, W., and Webster, K. (eds). 2019. *Proceedings of the First Workshop on Gender Bias in Natural Language Processing*. Florence: Association for Computational Linguistics.

Devlin, J., Chang, M.-W., Lee, K., and Toutanova, K. 2018. Bert: Pre-training of deep bidirectional transformers for language understanding. *arXiv preprint arXiv:1810.04805*.

Fokkens, A., Braake, S. T., Ockeloen, N., Vossen, P., Legêne, S., Schreiber, G., and de Boer, V. 2017a. BiographyNet: Extracting relations between people and events. *Europa baut auf Biographien (NAP 2017; Vol. 1)*., 193–227.

Fokkens, A., Ruigrok, N., Beukeboom, C., Sarah, G., and Van Atteveldt, W. 2018. Studying Muslim stereotyping through microportrait extraction. Pages 3734–3741 of: *Proceedings of the Eleventh International Conference on Language Resources and Evaluation*. LREC '18.

Fokkens, A., Ter Braake, S., Maks, I., Ceolin, D. 2016. On the semantics of concept drift: Towards formal definitions of semantic change. Pages 247–265 of: *Proceedings of Drift-a-LOD*. Bologna, Italy: CEUR Workshop Proceedings.

Fokkens, A., Van Erp, M., Vossen, P., Tonelli, S., Van Hage, W. R., Serafini, L., Sprugnoli, R., and Hoeksema, J. 2013. GAF: A grounded annotation framework for events. Pages 11–20 of: *Workshop on Events: Definition, Detection, Coreference, and Representation*.

Fokkens, A., Vossen, P., Rospocher, M., Hoekstra, R., van Hage, W. R., and Kessler, F. B. 2017b. GRaSP: Grounded representation and source perspective. *Proceedings of Knowledge Resources for the Socio-Economic Sciences and Humanities associated with RANLP*, **17**, 19–25.

Groth, P., Gibson, A., and Velterop, J. 2010. The anatomy of a nanopublication. *Information Services & Use*, **30**(1–2), 51–56.

Hellmann, S., Lehmann, J., Auer, S., and Brümmer, M. 2013. Integrating NLP using linked data. Pages 98–113 of: *Proceedings of the International Semantic Web Conference*. Berlin: Springer. `http://persistence.uni-leipzig.org/nlp2rdf/`.

Herbelot, A., Von Redecker, E., and Müller, J. 2012. Distributional techniques for philosophical enquiry. Pages 45–54 of: *Proceedings of the 6th Workshop on Language Technology for Cultural Heritage, Social Sciences, and Humanities*.

Lebo, T., Sahoo, S., McGuinness, D., Belhajjame, K., Cheney, J., Corsar, D., Garijo, D., Soiland-Reyes, S., Zednik, S., and Zhao, J. 2013. Prov-o: The prov ontology `http://www.w3.org/TR/prov-o/`.

Morante, R., Van Son, C., Maks, I., and Vossen, P. 2020. Annotating perspectives on vaccination. Pages 4964–4973 of: *Proceedings of the 12th Language Resources and Evaluation Conference*.

Moreau, L., Missier, P., Belhajjame, K., B'Far, R., Cheney, J., Coppens, S., Cresswell, S., Gil, Y., Groth, P., Klyne, G. 2013. Prov-dm: The prov data model. `https://www.w3.org/TR/prov-dm/`.

Ockeloen, N., Fokkens, A., Ter Braake, S., Vossen, P., De Boer, V., Schreiber, G., and Legêne, S. 2013. Biographynet: Managing provenance at multiple levels and from different perspectives. Pages 59–71 of: *Proceedings of the Workshop on Linked Science at International Semantic Web Conference 2013 (Vol. 1116)*. Sydney, Australia: CEUR-WS.org.

Peters, M. E., Neumann, M., Iyyer, M., Gardner, M., Clark, C., Lee, K., and Zettlemoyer, L. 2018. Deep contextualized word representations. *arXiv preprint arXiv:1802.05365*.

Shannon, C. E. 1948. A mathematical theory of communication. *The Bell System Technical Journal*, **27**(3), 379–423.

van Son, C., Caselli, T., Fokkens, A., Maks, I., Morante, R., Aroyo, L., and Vossen, P. 2016 (May). GRaSP: A multilayered annotation scheme for perspectives. Pages 1177–1184 of: *Tenth International Conference on Language Resources and Evaluation (LREC 2016)*. Paris, France: European Language Resources Association.

van Hage, W. R., Malaisé, V., Segers, R., Hollink, L., and Schreiber, G. 2011. Design and use of the simple event model (SEM). *Journal of Web Semantics*, **9**(2), 128–136.

Vossen, P., Agerri, R., Aldabe, I., Cybulska, A., van Erp, M., Fokkens, A., Laparra, E., Minard, A.-L., Aprosio, A. P., Rigau, G., Rospocher, M., and Segers, R. 2016. Newsreader: Using knowledge resources in a cross-lingual reading machine to generate more knowledge from massive streams of news. *Knowledge-Based Systems*, **110**, 60–85.

Vossen, P., Baez, S., Bajčetić, L., and Kraaijeveld, B. 2018. Leolani: A reference machine with a theory of mind for social communication. Pages 15–25 of: *International Conference on Text, Speech, and Dialogue*. Springer.

Vossen, P., Báez Santamaria, S., Bajčetić, L., Basić, S., and Kraaijeveld, B. 2019a. A communicative robot to learn about us and the world. *Proceedings of the Russian Conference on Computational Linguistics Dialogue 2019*.

Vossen, P., Baez, S., Bajčetić, L., Bašić, S., and Kraaijeveld, B. 2019b. Leolani: A robot that communicates and learns about the shared world. Pages 181–184 of: *2019 ISWC Satellite Tracks (Posters and Demonstrations, Industry, and Outrageous Ideas), ISWC 2019-Satellites*. CEUR-WS.

Vossen, P., Bajčetić, L., Báez Santamaria, S., Basić, S., and Kraaijeveld, B. 2019c. Modelling context awareness for a situated semantic agent. Pages 238–252 of *Proceedings of 11th International and Interdisciplinary Conference on Modeling and Using Context, CONTEXT 2019*. Springer.

Vossen, P., and Fokkens, A. 2021. Information systems, big data and knowledge landscapes. Pages 249–268 of: *Navigating Digital Health Landscapes*. Springer.

CPSIA information can be obtained
at www.ICGtesting.com
Printed in the USA
LVHW081523020522
717704LV00003B/90